T0216355

Lecture Notes in Artificial Intelligence   8940

Subseries of Lecture Notes in Computer Science

LNAI Series Editors

Randy Goebel
University of Alberta, Edmonton, Canada
Yuzuru Tanaka
Hokkaido University, Sapporo, Japan
Wolfgang Wahlster
DFKI and Saarland University, Saarbrücken, Germany

LNAI Founding Series Editor

Joerg Siekmann
DFKI and Saarland University, Saarbrücken, Germany

More information about this series at http://www.springer.com/series/1244

Martin Atzmueller · Alvin Chin
Christoph Scholz · Christoph Trattner (Eds.)

Mining, Modeling, and Recommending 'Things' in Social Media

4th International Workshops
MUSE 2013, Prague, Czech Republic, September 23, 2013
and MSM 2013, Paris, France, May 1, 2013
Revised Selected Papers

 Springer

Editors

Martin Atzmueller
University of Kassel
Kassel
Germany

Alvin Chin
Microsoft
Beijing
China

Christoph Scholz
University of Kassel
Kassel
Germany

Christoph Trattner
Norwegian University of Science
    and Technology
Trondheim
Norway

ISSN 0302-9743   ISSN 1611-3349   (electronic)
Lecture Notes in Artificial Intelligence
ISBN 978-3-319-14722-2   ISBN 978-3-319-14723-9   (eBook)
DOI 10.1007/978-3-319-14723-9

Library of Congress Control Number: 2014959272

LNCS Sublibrary: SL7 – Artificial Intelligence

Springer Cham Heidelberg New York Dordrecht London
© Springer International Publishing Switzerland 2015
This work is subject to copyright. All rights are reserved by the Publisher, whether the whole or part of the material is concerned, specifically the rights of translation, reprinting, reuse of illustrations, recitation, broadcasting, reproduction on microfilms or in any other physical way, and transmission or information storage and retrieval, electronic adaptation, computer software, or by similar or dissimilar methodology now known or hereafter developed.
The use of general descriptive names, registered names, trademarks, service marks, etc. in this publication does not imply, even in the absence of a specific statement, that such names are exempt from the relevant protective laws and regulations and therefore free for general use.
The publisher, the authors and the editors are safe to assume that the advice and information in this book are believed to be true and accurate at the date of publication. Neither the publisher nor the authors or the editors give a warranty, express or implied, with respect to the material contained herein or for any errors or omissions that may have been made.

Printed on acid-free paper

Springer International Publishing AG Switzerland is part of Springer Science+Business Media
(www.springer.com)

Preface

The areas of ubiquitous and social computing are creating new environments that foster the social interaction of users in several dimensions. On the ubiquitous side, there are different small distributed devices and sensors. For social media and social web, there are a variety of social networking environments being implemented in an increasing number of social media applications. With these, ubiquitous and social environments are transcending many diverse domains and contexts, including events and activities in business and personal life.

Altogether, understanding and modeling ubiquitous (and) social systems requires novel approaches, methods, and techniques for their analysis. This book sets out to explore this area by presenting a number of current approaches and studies addressing selected aspects of this problem space. The individual contributions of this book focus on problems related to the mining, modeling, and recommendation in ubiquitous social media, i.e., integrating both ubiquitous data and social media. Methods for mining, modeling, and recommendation can then help to advance our understanding of the dynamics and structures inherent to the respective systems integrating and applying ubiquitous social media, as well as for engineering applications.

Specifically, this book focuses on the collective intelligence in ubiquitous and social environments and how this data can be exploited to generate predictive models that serve as a basis for recommender systems in those environments. In this context, we present work that tackles issues such as personalization in social streams, recommendations exploiting social and ubiquitous data, and efficient information processing in social systems. Furthermore, this book presents work dealing with the problem of mining patterns from ubiquitous social data, including mobility mining and exploratory methods for ubiquitous data analytics.

The papers presented in this book are revised and significantly extended versions of papers submitted to two related workshops: The 4th International Workshop on Mining Ubiquitous and Social Environments (MUSE 2013), which was held on September 23, 2013 in conjunction with the European Conference on Machine Learning and Principles and Practice of Knowledge Discovery in Databases (ECML-PKDD 2013) in Prague, Czech Republic, and the 4th International Workshop on Modeling Social Media (MSM 2013), which was held on May 1, 2013 in conjunction with ACM Hypertext in Paris, France. With respect to these two complementing workshop themes, the papers contained in this volume form a starting point for bridging the gap between the social and ubiquitous worlds. Both social media applications and ubiquitous systems benefit from modeling aspects, either at the system level, or for providing a sound data basis for further analysis and mining options. On the other hand, data analysis and data mining can provide novel insights into the user's behavior within social media systems, and thus similarly enhance and support modeling prospects.

Concerning the range of topics, we broadly consider two main themes: predictive modeling in ubiquitous social data and pattern mining in ubiquitous social data.

For the first main theme, we included four works focusing on recommendation aspects in ubiquitous social media. We present "Network Activity Feed: Finding Needles in a Haystack" by Shlomo Berkovsky and Jill Freyne providing an interesting survey of related work tackling the problem of personalization of social network news feeds. In "Refining Frequency-Based Tag Reuse Predictions by Means of Time and Semantic Context" by Dominik Kowald, Simone Kopeinik, Paul Seitlinger, Tobias Ley, Dietrich Albert, and Christoph Trattner a novel recommendation approach based on the ACT-R theory to predict the user's tags is presented. In another interesting work "Forgetting the Words but Remembering the Meaning: Modeling Forgetting in a Verbal and Semantic Tag Recommender" by Dominik Kowald, Paul Seitlinger, Simone Kopeinik, Tobias Ley, and Christoph Trattner show how semantic and verbal decay influence the tag prediction problem. The paper "Ontology-Enabled Access Control and Privacy Recommendations" by Marcel Heupel, Lars Fischer, Mohamed Bourimi, and Simon Scerri presents the relevant problem of privacy in social networks and recommendation. In "Utilizing Online Social Network and Location-Based Data to Recommend Products and Categories in Online Marketplaces" by Emanuel Lacic, Dominik Kowald, Lukas Eberhard, Christoph Trattner, Leandro Balby Marinho, and Denis Parra the authors describe how location-based and social network data can be exploited to recommend products and categories efficiently.

For the second main theme, we included three works focusing on the pattern mining aspect. The paper "Exploratory Subgroup Analytics on Ubiquitous Data" by Martin Atzmueller, Juergen Mueller, and Martin Becker presents a study on exploratory subgroup analytics to obtain interesting descriptive patterns in ubiquitous data. In "Predictability Analysis of Aperiodic and Periodic Model for Long-Term Human Mobility Using Ambient Sensors" by Danaipat Sodkomkham, Roberto Legaspi, Ken-ichi Fukui, Koichi Moriyama, Satoshi Kurihara, and Masayuki Numao the authors present an analysis of the effectiveness of periodic and aperiodic predictive models on human mobility data. Finally, the paper "Open Smartphone Data for Structured Mobility and Utilization Analysis in Ubiquitous Systems" by Nico Piatkowski, Jochen Streicher, Katharina Morik, and Olaf Spinczyk sets out to present an open smartphone utilization and mobility dataset that was captured during a 4-month study period.

It is the hope of the editors that this book (i) catches the attention of an audience interested in recent problems and advancements in the fields of social media, online social networks, and ubiquitous data and (ii) helps to spark a conversation on new problems related to the engineering, modeling, mining, analysis, and recommendation in the field of ubiquitous social media and systems integrating these.

We want to thank the workshop and post-proceedings reviewers for their careful help in selecting and the authors for improving the submissions. We also thank all the authors for their contributions and the presenters for the interesting talks and the lively discussions at both workshops. Only this has allowed us to set up such a book.

November 2014   Martin Atzmueller
                                                                        Alvin Chin
                                                                  Christoph Scholz
                                                                Christoph Trattner

Organization

Program Committee

Martin Atzmueller	University of Kassel, Germany
Alejandro Bellogin	Centrum Wiskunde & Informatica, The Netherlands
Albert Bifet	University of Waikato, New Zealand
Robin Burke	DePaul University, USA
Javier Luis Canovas Izquierdo	Inria, École des Mines de Nantes, France
Ciro Cattuto	ISI Foundation, Italy
Michelangelo Ceci	Università degli Studi di Bari, Italy
Alvin Chin	Microsoft, China
Padraig Cunningham	University College Dublin, Ireland
Alexander Felfernig	Graz University of Technology, Austria
Michael Granitzer	University of Passau, Germany
Kris Jack	Mendeley, UK
Kristian Kersting	TU Dortmund University, Germany
Denis Parra	Pontificia Universidad Católica de Chile, Chile
Haggai Roitman	IBM Research Haifa, Israel
Christoph Scholz	University of Kassel, Germany
Philipp Singer	Knowledge Management Institute, Austria
Gerd Stumme	University of Kassel, Germany
Christoph Trattner	Norwegian University of Science and Technology, Norway
Zhiyong Yu	Fuzhou University, China
Arkaitz Zubiaga	Dublin Institute of Technology, Ireland

Additional Reviewer

Cosentino, Valerio

Contents

Exploratory Subgroup Analytics
on Ubiquitous Data

Martin Atzmueller[1]($\boxtimes$), Juergen Mueller[1], and Martin Becker[2]

[1] Knowledge and Data Engineering Group, University of Kassel, Kassel, Germany
{atzmueller,mueller}@cs.uni-kassel.de
[2] Data Mining and Information Retrieval Group,
University of Würzburg, Würzburg, Germany
becker@informatik.uni-wuerzburg.de

Abstract. This paper presents exploratory subgroup analytics on ubiquitous data: We propose subgroup discovery and assessment approaches for obtaining interesting descriptive patterns and provide a novel graph-based analysis approach for assessing the relations between the obtained subgroup set. This exploratory visualization approaches allows for the comparison of subgroups according to their relations to other subgroups and to include further parameters, e.g., geo-spatial distribution indicators. We present and discuss analysis results utilizing real-world data given by geo-tagged noise measurements with associated subjective perceptions and a set of tags describing the semantic context.

1 Introduction

Ubiquitous data mining has many facets including descriptive approaches: These can help for obtaining a first overview on a dataset, for summarization, for uncovering a set of interesting patterns, and analyzing their inter-relations.

In this paper, we propose exploratory subgroup analytics on ubiquitous data. Subgroup discovery is a versatile method for descriptive data mining. We extend it in two analytical directions focusing on the applied quality functions and the relations between subgroups. First, we propose novel quality functions for estimating the quality of subgroups in the multivariate setting. In addition, we propose a novel graph-based approach for assessing sets of subgroups including multiple quality criteria. Specifically, we focus on the interrelation between sensor measurements, subjective perceptions, and descriptive tags. For assessing the relations between the result set of subgroups, we propose a novel graph-based analysis approach. This method is applied for visualizing subgroup relations, and can be utilized for comparing subgroups according to their relationships to other subgroups. The automatic discovery and visual analysis methods complement each other in exploratory fashion: The quality function used for ranking and estimating the quality of subgroups and the relationship function can be selected according to analysis goals. In addition, further subgroup parameters can be shown in the visualization using visual markers. Here, we specifically present an adapted technique for deriving characteristic indicators of the geo-spatial distribution of ubiquitous data in the context of subgroup analytics.

© Springer International Publishing Switzerland 2015
M. Atzmueller et al. (Eds.): MUSE/MSM 2013, LNAI 8940, pp. 1–20, 2015.
DOI: 10.1007/978-3-319-14723-9_1

Overall, we analyze real-world sensor data with associated semantic information and subjective measurements. For that, we utilize the VIKAMINE[1] tool [8] for subgroup discovery and analytics; it is complemented by methods of the R environment for statistical computing [34] in order to implement a semi-automatic pattern discovery process.[2]

For the analysis, we apply real-world data from the EveryAware project:[3] Our application context is given by the *WideNoise* smartphone app for measuring environmental noise. The individual data points include the measured noise in decibel (dB), associated subjective perceptions (feeling, disturbance, isolation, and artificiality) and a set of tags for providing semantic context for the individual measurements. We present results analyzing subgroups patterns for hot-spots of low/high noise levels. Our results indicate, that there are indeed distinctive patterns in terms of descriptive tags. Furthermore, we analyze the characteristics of subgroups according to their geo-spatial distribution given by the covered set of noise measurements. In addition to investigating patterns that are characteristic for areas with low or high noise, we also analyze subgroups with respect to a distinctive perception profile – relating to subjective *perception patterns* – which we describe in terms of their assigned tags.

Our contribution can be summarized as follows:

1. We present an exploratory subgroup analytics approach covering the subgroup assessment and specifically the relations between subgroups. For this, we propose flexible quality functions and a set of relationship functions that are used to model dependencies and relations in a set of subgroups.
2. For the proposed automatic exploratory approach, we present a novel visualization method complementing the automatic methods. The presented visualization method allows to inspect subgroup relations and further influence parameters in context.
3. We provide an adapted technique for deriving geo-spatial distributional indicators in the context of subgroup analytics.
4. Finally, we present an analysis of ubiquitous data using subgroup discovery methods utilizing data from a real-world application. We describe a case study applying the proposed approaches and discuss the results in detail.

The remainder of the paper is organized as follows: Sect. 2 discusses related work. Then, Sect. 3 introduces necessary basic notions. Next, Sect. 4 proposes the novel approach for graph-based subgroup analytics. After that, Sect. 5 describes the applied dataset, presents the experiments, and discusses the results in our application setting. Finally, Sect. 6 concludes with a summary and presents interesting options for future work.

2   Related Work

Ubiquitous data mining covers many subfields, including spatio-temporal data mining [27], mining sensor data or mining social media with geo-referenced data,

---

[1] http://vikamine.org.
[2] http://rsubgroup.org.
[3] http://everyaware.eu.

c.f., [4]. Applications include destination recommenders, e.g., for tourist information systems [17], or geographical topic discovery [43]. Often established problem statements and methods have been transferred to this setting, for example, considering association rules [3]. Related approaches consider, for example, social image mining methods, cf., [33] for a survey.

In this area, specifically considering social image data, there have been several approaches, and the problem of generating representative tags for a given set of images is an active research topic. Reference [40] analyze Flickr data and provide a characterization on how users apply tags and which information is contained in the tag assignments. Their approach is embedded into a recommendation method for photo tagging, similar to [32] who analyze different aspects and contexts of the tag and image data. Reference [1] present a method to identify landmark photos using tags and social Flickr groups. They apply group information and statistical preprocessing of the tags for obtaining interesting landmark photos.

The concept of collecting information in ubiquitous systems, especially for crowd-sourced and citizen-driven applications is discussed in [36]. Basic issues of measuring noise pollution using mobile phones are presented in [38]. From a distributional point of view, the proposed approach seamlessly generalizes similar ones for analyzing event and place semantics using a user-specified quality function. Then we can capture techniques, e.g., for burst detection [24,35] or the analysis of peaks of temporal and spatial distributions [44]. Furthermore, such techniques can be incorporated in our exploratory analytics approach using visualization techniques.

In contrast to the approaches discussed above, this paper focuses on descriptive patterns. This allows for the flexible adaptation to the preferences of the users, since their interestingness can be flexibly tuned by selecting an appropriate quality function and target concept. There are several variants of pattern mining techniques, e.g., frequent pattern mining [20], mining association rules [2,28], and closed representations [16] as well as subgroup discovery [12,25,42], which is the method applied in this work. This work also extends existing subgroup discovery methods [9] for analyzing geo-tagged social media, especially in the direction of handling arbitrary sets of target properties and visual analytics. The latter methods are directly integrated into the subgroup analytics approach, for a holistic setting, similar to such visualizations for community mining [37].

For analyzing a set of subgroups, these are typically clustered according to their similarity, e.g., [10], or based on their predictive power [26]. Other methods for pattern set refinement and selection, e.g., [29] focus on similarities on the instance and/or description level. In contrast to these approaches, the proposed approach for subgroup set analytics generalizes those methods. We provide a general approach for analyzing subgroup relations based on a freely configurable "relationship" function, embedded in a graph-based framework for the assessment of sets of subgroups.

3 Preliminaries

Data mining includes descriptive and predictive approaches [21]. In the following, we focus on descriptive pattern mining methods. We apply subgroup discovery [25], a broadly applicable data mining method which aims at identifying interesting patterns with respect to a given target property of interest according to a specific quality function. This section first introduces the necessary notions concerning the data representation, subgroups and patterns, basics on graphs, and similarity measures.

3.1 Patterns and Subgroups

Below, we summarize basic notions on patterns and subgroups. We define subgroups and their descriptions as well as interestingness measures that are necessary in the context of exploratory subgroup analytics on ubiquitous data. For a more general view and a detailed discussion, we refer to, e.g., [9, 12].

Basic Definitions. Formally, a *database* $DB = (I, A)$ is given by a set of individuals I and a set of attributes A. A *selector* or *basic pattern* $sel_{a_i = v_j}$ is a Boolean function $I \rightarrow \{0, 1\}$ that is true if the value of attribute $a_i \in A$ is equal to v_j for the respective individual. The set of all basic patterns is denoted by S. For a numeric attribute a_{num} selectors $sel_{a_{num} \in [min_j; max_j]}$ can be defined analogously for each interval $[min_j; max_j]$ in the domain of a_{num}. The Boolean function is then set to true if the value of attribute a_{num} is within the respective range.

A *subgroup description* or (complex) *pattern sd* is then given by a set of basic patterns $sd = \{sel_1, \ldots, sel_l\}$, where $sel_i \in S$, which is interpreted as a conjunction, i.e., $sd(I) = sel_1 \wedge \ldots \wedge sel_l$, with $length(sd) = l$.

Without loss of generality, we focus on a conjunctive pattern language using nominal attribute–value pairs as defined above in this paper; internal disjunctions can also be generated by appropriate attribute–value construction methods, if necessary. A *subgroup (extension)*

$$sg_{sd} := ext(sd) := \{i \in I | sd(i) = true\}$$

is the set of all individuals which are covered by the pattern sd. As search space for subgroup discovery the set of all possible patterns 2^S is used, that is, all combinations of the basic patterns contained in S.

Interestingness of a Pattern. A *quality function* $q: 2^S \rightarrow \mathbb{R}$ maps every pattern in the search space to a real number that reflects the interestingness of a pattern (or the extension of the pattern, respectively).

The result of a subgroup discovery task is the set of k (subgroup) patterns $res_1, \ldots, res_k$, where $res_i \in 2^S$ with the highest interestingness according to the quality function. While a large number of quality functions has been proposed in literature, many quality measures trade-off the size $n = |ext(sd)|$ of a subgroup and the deviation $t_{sd} - t_0$, where t_{sd} is the average value of a given target concept in the subgroup identified by the pattern sd and t_0 the average value of the target concept in the general population.

Thus, typical quality functions are of the form

$$q_a(sd) = n^a \cdot (t_{sd} - t_0),\ a \in [0; 1]\,. \tag{1}$$

For binary target concepts, this includes, for example, the *weighted relative accuracy* for the size parameter $a = 1$ or a simplified binomial function, for $a = 0.5$.

An extension to a target concept defined by a set of variables can be defined similarly, by extending common statistical tests. For comparing multivariate means for a set of m numeric attributes T_M, with $m = |T_M|$, for example, we can make use of Hotelling's T-squared test [22], for the quality measure q_H:

$$q_H = \frac{n(n-m)}{m(n-1)} (\mu_P^{T_M} - \mu_0^{T_M})^\top CV_P^{T_M\,-1} (\mu_P^{T_M} - \mu_0^{T_M})\,,$$

where $\mu_P^{T_M}$ is the vector of the model attribute means in the subgroup sg_{sd}, $CV_P^{T_M}$ is the covariance matrix, and $\mu_0^{T_M}$ is the vector of the target concept means in DB.

3.2 Distributional Geo-Spatial Subgroup Analysis

For analyzing the geo-spatial distribution of a subgroup, we adapt the approach presented in [44], which we briefly summarize below. We are basically interested in the geographical distribution of a subgroup, and in a characterization of this distribution using an adequate measure. Specifically, we model the geo-spatial distribution of a subgroup, and derive a *peakiness* measure of this distribution, indicating the overall shape of the distribution.

For modeling the distribution, first the geo-space of the world is split into b bins by r_{lat} degrees of latitude and r_{lon} degrees of longitude. Then, the latitude and longitude coordinates of the individuals $i \in sg_{sd}$ of the pattern sd can be mapped into those bins. We obtain a vector $v(sd) = (v_1, \dots, v_b)$ of occurrence counts v_i of the subgroup pattern for each bin $i, i = 1 \dots b$, i.e., how many distinct users have used the specific pattern sd there. We normalize this vector, by dividing each entry by the sum of the absolute values of the vector's entries (L1-Norm $\|\cdot\|_1$). Finally, we obtain the *peakiness* value $\phi(sd)$ of a pattern utilizing the vector $v = v(sd)$ by computing its second moment, as follows:

$$\phi(sd) = \frac{v \cdot v}{\|v\|_1^2} = \frac{1}{\|v\|_1^2} \sum_{i=1}^{b} v_i^2 \tag{2}$$

A high *peakiness* indicates rather characteristically peaky distributions, while a low value is observed for distributions close to a uniform distribution, cf. [44].

3.3 Graphs

An (undirected) *graph* $G = (V, E)$ is an ordered pair, consisting of a finite set V containing the *vertices* (also called *nodes*), and a set E of *edges* denoting the *connections* between the vertices. In the following, we freely use the term *network* as a synonym for the term graph. A *weighted* graph is a graph $G = (V, E)$ together with a function w : $E \to \mathbb{R}^+$ that assigns a positive weight to each edge. The *density* of G is the fraction of possible edges that are actually present. The *degree* d(u) of a node u in a network measures the number of connections it has to other nodes. In weighted graphs the *strength* s(u) is the sum of the weights of all edges containing u, i.e., s(u) := $\sum_{\{u,v\} \in E} w(\{u, v\})$. A (weakly) connected component of G is a subset $U \subseteq V$, such that there exists an (undirected) path between every pair of nodes $\{u, v\}, u, v \in U$, i.e., that u and v are connected by a sequence of edges. For more details, we refer to standard literature, e.g., [18].

3.4   Similarity Measures

Given two vectors $\boldsymbol{v}_1, \boldsymbol{v}_2 \in \mathbb{R}^X$, there are a variety of *similarity measures* for assessing the similarity between the contained values, e.g., [23,41] for a detailed overview. We can measure a *Manhattan similarity*, for example, by utilizing the (normalized) Manhattan distance, defined as follows:

$$\mathrm{sim}_{\mathrm{man}}(\boldsymbol{v}_1, \boldsymbol{v}_2) := 1 - \frac{\sum_{i=1}^{X} |\boldsymbol{v}_{1i} - \boldsymbol{v}_{2i}|}{X}, \tag{3}$$

where v_{ij} denotes the j-th component of vector v_i.

An alternative measure known from information retrieval is the cosine measure. The *cosine similarity* between two vectors $\boldsymbol{v}_1, \boldsymbol{v}_2 \in \mathbb{R}^X$ is then defined as:

$$\mathrm{sim}_{\mathrm{cos}}(\boldsymbol{v}_1, \boldsymbol{v}_2) := \cos \angle(\boldsymbol{v}_1, \boldsymbol{v}_2) = \frac{\boldsymbol{v}_1 \cdot \boldsymbol{v}_2}{||\boldsymbol{v}_1||_2 \cdot ||\boldsymbol{v}_2||_2}. \tag{4}$$

4   Exploratory Subgroup Analytics

In the following, we first present our novel subgroup analytics approach using a graph-based representation for inspecting and assessing a set of subgroups. We describe a set of analysis methods for exploratory data analysis using subgroup discovery, and provide automatic and interactive techniques. In summary, we propose the following methods for exploratory subgroup analytics that are described in more detail below.

- Exploratory ubiquitous subgroup analytics of noise and perception patterns.
- Subgroup-based analysis of spatio-temporal peakiness.
- Visual exploration and assessment of subgroup relations.

4.1   Exploratory Ubiquitous Subgroup Analytics

As a first step in our subgroup analytics approach, we obtain a set of the top-k subgroups for a specific target variable. Depending on the analytical questions, different quality functions can be applied, as sketched above, such that relatively simple deviations of a target variable from the "overall trend" observed in the general population can be analyzed. Furthermore, more complex exceptionality criteria, e.g., corresponding to more complex models such as captured by a set of target variables can be investigated.

In an exhaustive subgroup discovery approach, all combinations of possible selectors (e.g., tags) are analyzed for discovering interesting patterns. Since this search space is exponential in the number of selectors, typically an efficient subgroup discovery algorithm needs to be applied, e.g., the BSD [31], SD-Map [12], or the SD-Map* [7] algorithm. For large ubiquitous data with sparse distributions, the SD-Map* algorithm can often be successfully applied, cf. [9].

The result of the subgroup discovery step is then given by the top-k subgroups that need to be assessed and put into relation to each other. This is typically performed semi-automatically, based on automatic methods guided by user interaction. Then, the result set of subgroups can be inspected and validated. This can be supported by background knowledge [11,14], statistical approaches [8] and interactive techniques, e.g., [10] that are applied according to the *Information Seeking Mantra* by Shneiderman [39]: Overview first (macroscopic view), browsing and zooming (mesoscopic analysis), and details on demand (microscopic focus).

4.2 Subgroup-Based Analysis of Spatio-Temporal Peakiness

Basic subgroup discovery described in the previous section provides a first view on interesting patterns and data characteristics for selected target concepts. However, ubiquitous data typically also often contains temporal and specifically geo-spatial information. The latter is especially interesting for identifying, e.g., interesting places and locations.

In order to analyze the data further into this direction, we apply the technique described in Sect. 3 concerning the analysis of peakiness of subgroup patterns. *Peaky* subgroups are then those which are relatively specific for a certain *location* considering the geo-spatial information. Essentially, this information can be computed in addition to the subgroup information, i. e., a peakiness value is assigned to each subgroup that we collected using subgroup discovery.

For computing the peakiness value, we utilize the parameters proposed in [44] and divide the space into a one-by-one degree grid, and use the latitude and longitude values associated with our ubiquitous data measurements. Then, for each subset of the database covered by a specific subgroup, we can determine the peakiness value. These can then be added to the subgroup information and furthermore visualized in several ways as presented in the case study below.

4.3 Visual Exploration and Assessment of Subgroup Relations

After subgroups and their parameters, e.g., their peakiness values, have been determined, we apply an interative step for the visual exploration and assessment of a set of subgroups. Using a given relationship function, we consider specific *relations* between subgroups. Then, their "connections" according to this relation can be modeled as a graph.

More formally, given a certain criterion implemented by a relation function $rel : I \times I \to \mathbb{R}$ we obtain a value estimating the relationship between pairs of subgroups, identified by their respective subgroup descriptions. Possible relations include, for example, geographic distance, or semantic criteria. In our application setting, we focus on the latter, since we will use the given perceptions for noise measurements as semantic proxies for subgroup relatedness.

For assessing our result set of subgroups R, we obtain the rel-value for each pair of subgroups (u, v). After that, we construct a *subgroup assessment graph* G_R for R: The nodes of G_R are given by the subgroups contained in R. The edges between node pairs (u, v) are constructed according to the respective $rel(u, v)$ value: If the respective value between the subgroup pair is zero, then the edge is dropped; otherwise, an edge weighted by $rel(u, v)$ is added to the graph. In addition to the connections denoted by edges in the graph, we can furthermore visualize certain parameters such as properties of the nodes by colors and/or size of the nodes, as well as the weights of edges by different edge styles such as thickness or line types. In the graph, we can directly visualize the connectedness of a node using the degree information, as well as the peakiness in order to directly highlight highly interesting subgroups.

It is easy to see that, depending on the applied relationship function rel, the graph construction process can result in a fully connected graph which is hard to interpret. Therefore, a refinement of this process utilizes a certain threshold τ_{rel} which is used for pruning edges in the graph. If the relation "strength" $rel(u, v)$ between a subgroup pair (u, v) is below the threshold, i.e., $rel(u, v) < \tau_{rel}$ then we do not consider the edge between u and v, such that the edge is dropped. By carefully selecting a suitable threshold τ_{rel} the resulting subgroup network can then be easily inspected and assessed.

Typically, the situation becomes interesting when the graph is split into different components corresponding to certain clusters of subgroups. We will discuss examples of constructed networks below. For selecting a suitable threshold, a *threshold-component* visualization can be applied, see Fig. 10 for an example. This visualization plots the number of connected components of the graph depending on the applied threshold. Then, the "steps" within the plot can indicate interesting thresholds that can be interactively inspected. A related visualization plots the used threshold against the graph density for obtaining a first impression of the ranges of suitable threshold selections, cf. Fig. 11.

5   *WideNoise Plus* Case Study

In the following, we first describe the applied dataset. Then, we discuss a basic statistical analysis of the main parameters concerning the noise levels and tag distribution. After that, we describe a case study applying the presented techniques and discuss our results in context.

5.1 Applied Dataset

In this paper, we utilize data from the EveryAware project, specifically, on collectively organized noise measurements collected using the *WideNoise Plus* application between December 14, 2011 and June 6, 2014.

WideNoise Plus allows the collection of noise measurements using smartphones. It includes sensor data from the microphone given as noise level in dB(A), the location from the GPS-, GSM-, and WLAN-sensor represented as latitude and longitude coordinate, as well as a timestamp. Furthermore, the user can enter his perceptions about the measurement, expressed using the four sliders for feeling (love to hate), disturbance (calm to hectic), isolation (alone to social), and artificiality (nature to man-made). In addition, tags can be assigned to the recording. We collected data from all around the world using iOS and Android devices. The largest user group is located around Heathrow Airport London where the residents map the noise pollution caused by the airport to profile and monitor their environmental situation.

The data are stored and processed using the EveryAware backend [15], which is based on the UBICON software platform [5,6].

The applied dataset contains 6,600 data records and 2,009 distinct tags: The available tagging information was cleaned such that only tags with a length of at least three characters were considered. Only data records with valid tag assignments were included. Furthermore, we applied stemming and split multiword tags into distinct single word tags. In our analysis, we utilize the following objective and subjective information for each measurement:

- Objective: Level of noise (dB).
- Subjective perceptions about the environment, encoded in the interval $[-5; 5]$:
  - "Feeling" (hate/love) where -5 is most extreme for "hate" and 5 is most extreme for "love".
  - "Disturbance" (hectic/calm) where -5 is most extreme for "hectic" and 5 is most extreme for "calm".
  - "Isolation" (alone/social) where -5 is most extreme for "alone" and 5 is most extreme for "social".
  - "Artificiality" (man-made/nature) where -5 is most extreme for "man-made" and 5 is most extreme for "nature".
- Tags, e.g., "noisy", "indoor", or "calm", providing the semantic context of the specific measurement.

Figures 1, 2, 3 and 4 show the value distributions of the different perception values as histograms.

5.2 Statistical Analysis

In this section, we perform some basic statistical analysis of the observed distributions as well as initial experiments on correlating the subjective and objective data. As we will see, we observe typical phenomena in the domain of tagging data, while the correlations are expressed on a medium level; this directly leads

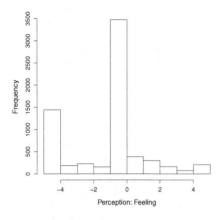

Fig. 1. Histogram for the subjective perception 'feeling'.

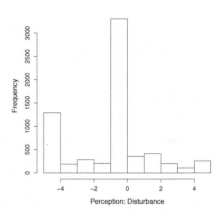

Fig. 2. Histogram for the subjective perception 'disturbance'.

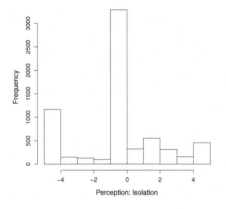

Fig. 3. Histogram for the subjective perception 'isolation'.

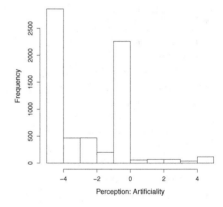

Fig. 4. Histogram for the subjective perception 'artificiality'.

to the advanced techniques using our subgroup discovery that we describe in the subsequent sections, where we analyze the relation between objective and subjective data given patterns of tagging data in more detail.

Figures 5, 6, 7 and 8 provide basic statistics about the tag count and measured noise distributions, as well as the value distributions of the perceptions and the number of tags assigned to a measurement. Figure 6 shows the distribution of the collected dB values, with a mean of 67.42 dB.

In Fig. 7 we observe a typical heavy-tailed distributions of the tag assignments. Also, as can be observed in Figs. 5 and 8, the tag assignment data is rather sparse, especially concerning larger sets of assigned tags. However, it already allows to draw some conclusions on the tagging semantics and perceptions.

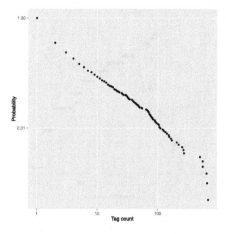

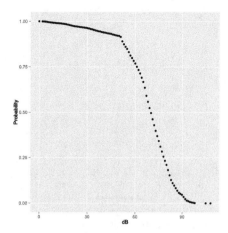

Fig. 5. Cumulated tag count distribution in the dataset. The y-axis provides the probability of observing a tag count larger than a certain threshold on the x-axis.

Fig. 6. Cumulated distribution of noise measurement (dB). The y-axis provides the probability for observing a measurement with a dB value larger than a certain threshold on the x-axis.

Table 1. Correlation analysis between subjective (perceptions) and objective (dB) measurements; all values are statistically significant ($p < 0.01$).

	Feeling	Disturbance	Isolation	Artificiality
dB	-0.27	-0.32	-0.32	0.19

In this context, the relation between (subjective) perceptions and (objective) noise measurements is of special interest. Table 1 shows the results of analyzing the correlation between the subjective and objective data. As shown in the table, we observe the expected trend that higher noise values correlate with the subjective "hate", "hectic" or "man-made" situations. While the individual correlation values demonstrate only medium correlations, they are nevertheless statistically significant.

5.3 Exploratory Subgroup Analytics: Results and Discussion

In the following, we present exploratory subgroup discovery results in the context of the *WideNoise Plus* data. First, we focus on the analysis of the (subjective) perceptions as target concepts, for which we identify both highly deviating and "conforming" patterns, i.e., those that are close to the means of the perceptions observed in the complete dataset. After that, we analyze characteristic patterns for high and low noise levels. Finally, we present a combined analysis, also including the geo-spatial distribution of specific subgroups.

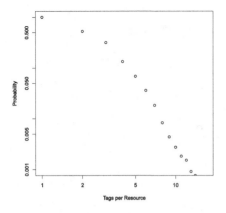

Fig. 7. Cumulated tag per record distribution in the dataset. The y-axis provides the probability of observing a tag per record count larger than a certain threshold on the x-axis.

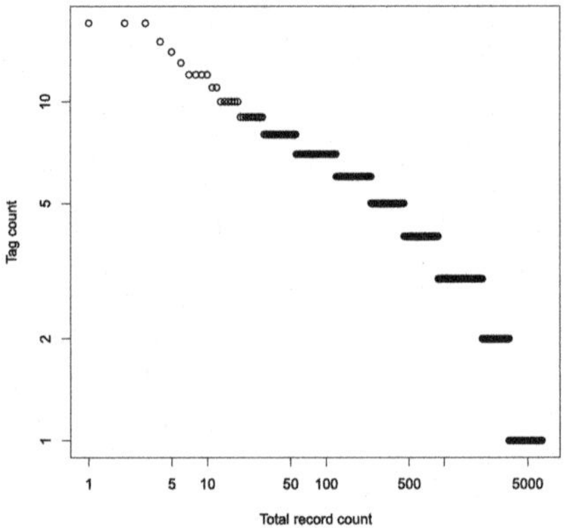

Fig. 8. Distribution of assigned tags per resource/data record.

Analysis of Perception Patterns. For analyzing the characteristics of the subjective data given by the perception values assigned to the individual measurements we applied the multi-target quality function q_H (based on the Hotelling's T-squared test); in that way, we measured, which patterns show a perception profile (given by the means of the individual perceptions) that is exceptionally different from the overall picture of the perceptions (means on the complete dataset). In addition, we also analyzed, which patterns show a rather "conforming" behavior to the overall mean values. For that we applied the quality function

$$q'_H = \frac{1}{q_H}.$$

Using the reciprocal of q_H we could then identify patterns for which their deviation was quite small, i.e., close to the general trend in the complete dataset. Table 2 presents the obtained results, where the rows 1–10 in the table denote deviating patterns (q_H), while rows 11–20 show conforming patterns (q'_H).

For comparison, the overall means of the perceptions are given by: feeling = −0.83, disturbance = −0.64, isolation = −0.19, artificiality = −2.33. As we can observe in the table, the deviating patterns tend to correspond to more *noisy* patterns; the majority of the patterns shows a dB value above the mean in the complete dataset (67.42 dB). Furthermore, most of the patterns relate to the Heathrow case study, e.g., *north AND runway, plane AND south*; an interesting pattern is given by *plane AND runway AND garden* – residents close to Heathrow obviously tend to measure noise in their garden. For the *conforming* patterns we mostly observe patterns with a mean dB close to the general mean. However, interestingly there are some patterns that show an increased mean and also "unexpected" patterns, e.g., *street AND traffic* or *airport*.

Table 2. Perception patterns: rows 1–10 - deviating patterns; rows 11–20 - conforming patterns; Overall means (perceptions): feeling $= -0.83$, disturbance $= -0.64$, isolation $= -0.19$, artificiality $= -2.33$. The table shows the size of the subgroups, their quality according to the applied quality function, the mean of the measured dB values, and the means of the individual perceptions.

id	description	size	quality	mean dB	feeling	disturbance	isolation	artificiality
1	north AND runway	31	6223.79	80.32	-4.87	-4.97	-4.32	-4.97
2	heathrow	635	3609.66	69.71	-4.84	-4.79	-4.21	-4.90
3	aeroplan	550	3345.64	67.29	-4.79	-4.71	-4.70	-4.79
4	north	32	1813.34	79.59	-4.69	-4.69	-4.31	-4.97
5	esterno	548	1660.91	69.86	0.99	1.34	1.55	-1.89
6	plane AND runway AND garden	33	1237.88	79.45	-2.21	-2.27	1.09	-2.24
7	nois	648	1214.25	66.34	-4.39	-4.14	-4.20	-4.29
8	plane AND south	65	1186.62	79.54	-3.29	-3.12	-0.35	-3.29
9	voci	270	1138.21	71.80	0.93	1.32	2.10	-2.32
10	plane AND runway	91	999.63	79.96	-3.74	-3.66	-1.45	-3.77
11	park	26	0.72	66.69	-0.19	0.12	-0.81	-0.85
12	san	27	0.50	70.74	-0.15	-0.22	0.04	-1.37
13	lorenzo AND outdoor	22	0.29	70.77	0.00	-0.14	0.32	-1.27
14	street AND traffic	33	0.25	70.12	-1.55	-0.88	0.61	-3.45
15	univers	25	0.24	57.20	-0.32	0.32	0.88	-2.16
16	lorenzo	25	0.23	71.00	0.04	0.00	0.32	-1.16
17	land AND nois	20	0.20	75.80	-2.70	-1.15	0.10	-1.65
18	work	92	0.20	56.27	-0.40	0.23	-0.32	-1.67
19	room	25	0.19	50.52	1.08	1.36	-1.16	-1.96
20	airport	23	0.17	72.57	-0.04	-1.35	1.96	-3.26

Overall, these results confirm the trends that we observed in the statistical analysis above indicating a medium correlation of the perceptions with the noise patterns. While the analysis of the perceptions provides some initial insights on subjective and objective data, again these results motivate our proposed approach for analyzing subgroups and their relations modeled by arbitrary parameters in more detail. This will be discussed in the next section, where we provide an integrated approach for assessing noise and perceptions patterns and their inter-relations.

Analysis of Noise (dB) Patterns. For identifying characteristic noise patterns, we applied subgroup discovery for the target variable *noise (dB)* focusing on subgroups both with a large deviation comparing the mean of the target in the subgroup and the target in the complete database. It is easy to see that increasing deviations (above the global mean) indicate *noisy* environments, while decreasing deviations (below the global mean) indicate more *quiet* situations. For analysis, we applied the simple binominal quality function, c.f., Sect. 3 and discovered the top-20 patterns for *noisy* and *quiet* environments, respectively.

Table 3 shows 40 patterns combining the two top-20 result sets. The patterns in rows $1-20$ denote the top-20 patterns for the target concept "high dB Value" whereas the subgroup patterns in rows $21-40$ denote the top-20 patterns for the target concept "low dB Value".

Table 3. Patterns: rows 1–20 - target: large mean noise (dB); rows 21–40 - target: small mean noise (dB); Overall mean (population): 67.42 dB. The last two columns include the node degree in the subgroup assessment graph, for $\tau_{rel} = 0.90$ and $\tau_{rel} = 0.95$.

id	description	size	mean dB	feeling	disturbance	isolation	artificiality	peakiness	deg (t=0.90)	deg (t=0.95)
1	craft	66	92.14	-3.03	-3.18	3.18	-4.61	1.00	9	2
2	air	75	88.43	-3.03	-3.09	2.73	-4.49	0.85	9	2
3	arriva	252	78.64	-0.02	-0.01	0.01	0.00	0.99	23	19
4	plane	495	73.88	-3.40	-2.25	-0.53	-3.38	0.97	8	5
5	aircraft	154	77.44	-1.18	-0.31	-0.71	-2.81	0.85	10	4
6	garden	674	72.19	-0.32	-0.17	-0.20	-4.39	0.90	10	4
7	plane AND runway	91	79.96	-3.74	-3.66	-1.45	-3.77	0.98	7	2
8	runway	100	79.12	-3.70	-3.62	-1.60	-3.88	0.98	7	2
9	heathrow AND plane	33	86.79	-4.33	-4.21	-0.30	-4.36	0.94	8	2
10	aeroporto	15	94.00	-5.00	0.00	0.00	0.00	1.00	10	4
11	ciampino	18	91.39	-4.17	0.00	0.00	0.00	1.00	17	5
12	plane AND south	65	79.54	-3.29	-3.12	-0.35	-3.29	0.97	8	3
13	runway AND south	65	79.54	-3.29	-3.12	-0.35	-3.29	0.97	8	3
14	south	66	79.41	-3.24	-3.08	-0.35	-3.32	0.97	8	3
15	plane AND street	46	80.83	-4.85	-3.87	-2.35	-4.87	1.00	4	1
16	ferri	20	86.95	0.05	0.05	0.10	-0.10	0.59	23	19
17	runway AND street	40	80.68	-4.83	-4.58	-2.83	-4.85	1.00	2	1
18	rout	120	75.07	-0.04	-0.02	0.03	-0.03	1.00	23	19
19	bus	170	73.83	-0.93	-1.24	0.76	-2.16	0.17	23	6
20	eva	71	77.11	0.00	0.00	0.00	-0.07	0.89	23	19
21	home	164	43.71	1.15	1.31	-0.99	-0.94	0.23	22	19
22	background	79	49.84	0.15	1.61	-1.08	-0.49	0.50	23	19
23	indoor	151	54.78	0.60	0.56	-0.08	-1.19	0.21	23	19
24	bosco	14	33.36	3.21	3.36	-1.71	1.93	1.00	0	0
25	night	32	45.53	1.78	2.38	-1.59	0.00	0.09	20	1
26	offic	204	58.77	0.08	0.73	-0.46	-1.69	0.22	27	20
27	fan AND music	14	37.50	0.21	0.29	0.00	-0.29	0.87	23	19
28	background AND nois	37	49.14	0.32	1.95	-2.30	-1.05	1.00	27	20
29	fan	23	44.35	-0.26	0.22	-1.30	-1.48	0.39	25	6
30	general	42	50.45	0.00	1.31	0.00	0.00	1.00	21	15
31	fan AND indoor	15	39.40	0.13	0.33	-0.07	-0.33	0.76	23	19
32	work	92	56.27	-0.40	0.23	-0.32	-1.67	0.33	26	11
33	morn	30	47.93	0.43	1.80	-1.93	-1.20	0.60	24	20
34	background AND indoor	23	45.35	0.43	1.74	-2.00	-0.91	1.00	24	19
35	indoor AND nois	23	45.35	0.43	1.74	-2.00	-0.91	1.00	24	19
36	background AND work	23	47.43	0.61	1.74	-2.00	-0.74	1.00	24	20
37	nois AND work	23	47.43	0.61	1.74	-2.00	-0.74	1.00	24	20
38	kassel	110	58.35	-0.05	0.54	0.12	-1.27	0.91	24	20
39	room	25	50.52	1.08	1.36	-1.16	-1.96	0.07	24	20
40	indoor AND music	20	48.65	0.60	0.15	0.50	-0.70	0.55	22	19

In the table, we can identify several distinctive tags for noisy environments, for example, *north AND runway*, *heathrow*, and *aeroplan*, which relate to Heathrow noise monitoring, c.f., [5] for more details. These results confirm the results of the basic analysis in [5]. For more quiet environments, we can also observe typical patterns, e.g., focusing on the tags *park*, *lorenzo AND outdoor*, and *room*, and combinations. Some further interesting subgroups are described by the tags *bosco* (forest) and *night*. These also show a quit distinct perception profile, shown in the respective columns of Table 3. This can also be observed in the last two columns of the table indicating the degree in the subgroup assessment graph (see below): The subgroups described by *bosco* and *night* are quite isolated.

When considering the *peakiness* of the subgroups, we observe that most of the patterns are relatively specific for certain locations, since they exhibit rather high peakiness values, e.g., *craft*, *plane*, *aeroporto* etc. These correspond to relatively

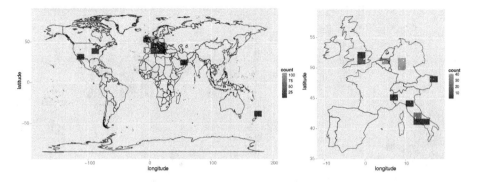

Fig. 9. Exemplary peakiness visualization for the pattern *bus* (worldwide, Europe).

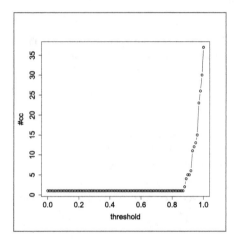
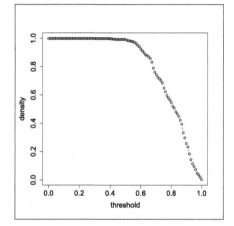

Fig. 10. Thresholded connected component plot based on a minimal *rel* value.

Fig. 11. Thresholded density plot based on a minimal *rel* value.

specific terms indicating certain locations. In contrast, there are some tags for which we estimate rather low peakiness values, for example, *bus*, *indoor*, *night*, *offic(e)*, *work*, *room*. It is easy to see, that those tags correspond to more *general* terms that are not so specific for certain locations. Therefore, these results are first a first indication for the relevance of the peakiness indicator with respect to identifying specific locations as described by specific tags. Figure 9 shows an example of visualizing the pattern *bus* on a worldwide view and a map of Europe, respectively. The subfigures show the location of the individual measurements, while the counts observed in different regions are color-coded: Lighter blue-areas indicate locations with higher counts of measurements. The figures are an example for a pattern with a smaller peakiness value, since the measurements are distributed more widely on a worldwide scale. In contrast, a pattern such as *kassel* tends to be more focused arund the kassel area (top-right light-blue area in the right plot of Fig. 9).

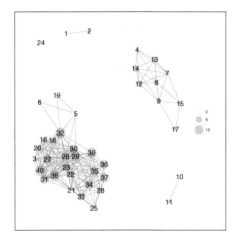

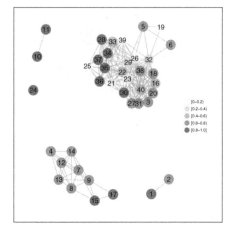

Fig. 12. Assessment graph: $\tau_{rel} = 0.90$; the size of a node depicts its degree.

Fig. 13. Assessment graph: $\tau_{rel} = 0.95$ with color-coded peakiness values.

Graph-Based Exploration of Subgroup Relations. In order to analyze subgroup relations with respect to the perceptions, we apply the Manhattan similarity as defined in Sect. 3 as our assessment relation *rel*. We measure the similarity using the averaged perception vectors of the respective subgroup patterns, with normalized values in the interval $[0; 1]$. Using the Manhattan similarity, we consider the overall "closeness" of the vectors; alternatively, the cosine similarity would focus on similar perception "profiles", i. e., uniformly expressed perceptions.

For determining appropriate thresholds τ_{rel}, Fig. 10 shows a threshold vs. connected component plot, constructed using the given similarity measure for estimating the relations between the subgroup patterns. Figure 11 shows the according thresholded density plot. Then, appropriate thresholds can be selected by the analyst. Figures 12 and 13 show the graphs for a threshold $\tau_{rel} = 0.9$, for which the degrees of the nodes are visualized by the size of the individual nodes, and the peakiness is color-coded, respectively.

As can be observed in Figs. 12 and 14 the respective networks for thresholds 0.90 and 0.95 show a distinct structure. Starting with $\tau_{rel} = 0.90$ the networks start to break up into distinct components, for $\tau_{rel} = 0.95$ the number of component increases significantly. For the lowest threshold $\tau_{rel} = 0.90$ we can already observe the special structure of pattern 24, one larger, and two smaller clusters. With threshold $\tau_{rel} = 0.95$, several more clusters emerge – the "Heathrow clusters" (7, 8), (15, 17) as well as the large cluster covering most of the *lower noise* patterns. However, this cluster also contains some patterns from the *higher noise* patterns (5, 6), which are rather unexpected and therefore quite interesting for subsequent analysis of the assigned perceptions. The connecting subgroup patterns can then be simply extracted by tracing the connections in the graph.

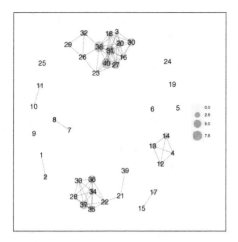
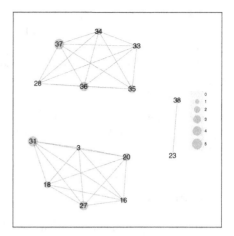
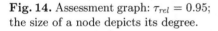

Fig. 14. Assessment graph: $\tau_{rel} = 0.95$; the size of a node depicts its degree.

Fig. 15. Assessment graph: $\tau_{rel} = 0.97$ zoomed in on the large cluster in Fig. 14; the size of a node depicts its degree.

Using the *Information Seeking Mantra*, cf. [39], we can zoom in and analyze details on demand as described above. Figure 15 shows an example, focusing on the high-degree nodes (degree ≥ 5) of the large cluster discussed above (Fig. 14), for a threshold $\tau_{rel} = 0.97$. The cluster dissolves into distinct components and the strongly connected (and partially overlapping subgroups) remain, for example, the patterns $34, 35, 36$.

6 Conclusions

In this paper, we presented exploratory subgroup analytics for obtaining interesting descriptive patterns in ubiquitous data. The presented approach includes semi-automatic techniques for comprehensive analysis of target concepts ranging from single variables to multi-target analysis and geo-spatial patterns. Specifically, we provided a novel graph-based analysis approach for assessing the relations between sets of subgroups including additional properties such as connectivity of the patterns or their peakiness corresponding to geo-spatial locations. Using data from a ubiquitous application we presented the proposed approach and discussed analysis results of a real-world case study. The analyzed noise measurements and associated subjective perceptions described by a set of tags confirmed the semantic context and provided interesting patterns with respect to the analysis of subjective and objective data.

For future work, we aim to extend the approach to diverse relationship and similarity measures. Furthermore, we plan to investigate multi-relational representations, i.e., multi-graphs capturing a set of relationships for assessing a set of subgroups. A further direction for analysis concerns the analyis of interrelations between perceptions, tags, and sentiments based on the tagging data, e.g., extending case-based approaches, e.g., [13] and methods for local exceptionality detection, e.g., [30]. These can then also be applied, for example, for enhanced event detection, recommendations, or community mining.

Acknowledgements. This work has been supported by the VENUS research cluster at the interdisciplinary Research Center for Information System Design (ITeG) at Kassel University, and parts of this research was funded by the European Union in the 7th Framework programme EveryAware project (FET-Open).

References

1. Abbasi, R., Chernov, S., Nejdl, W., Paiu, R., Staab, S.: Exploiting flickr tags and groups for finding landmark photos. In: Boughanem, M., Berrut, C., Mothe, J., Soule-Dupuy, C. (eds.) ECIR 2009. LNCS, vol. 5478, pp. 654–661. Springer, Heidelberg (2009)
2. Agrawal, R., Srikant, R.: Fast algorithms for mining association rules. In: Proceedings of VLDB, pp. 487–499. Morgan Kaufmann (1994)
3. Appice, A., Ceci, M., Lanza, A., Lisi, F., Malerba, D.: Discovery of spatial association rules in geo-referenced census data: a relational mining approach. Intell. Data Anal. **7**(6), 541–566 (2003)
4. Atzmueller, M.: Mining social media: key players, sentiments, and communities. WIREs: Data Min. Knowl. Disc. **2**(5), 411–419 (2012)
5. Atzmueller, M., Becker, M., Doerfel, S., Kibanov, M., Hotho, A., Macek, B.E., Mitzlaff, F., Mueller, J., Scholz, C., Stumme, G.: Ubicon: observing social and physical activities. In: Proceedings of IEEE International Conference on Cyber, Physical and Social Computing, pp. 317–324. IEEE Computer Society, Washington, DC, USA (2012)
6. Atzmueller, M., Becker, M., Kibanov, M., Scholz, C., Doerfel, S., Hotho, A., Macek, B.E., Mitzlaff, F., Mueller, J., Stumme, G.: Ubicon and its applications for ubiquitous social computing. N. Rev. Hypermedia Multimedia **20**(1), 53–77 (2014)
7. Atzmueller, M., Lemmerich, F.: Fast subgroup discovery for continuous target concepts. In: Rauch, J., Raś, Z.W., Berka, P., Elomaa, T. (eds.) ISMIS 2009. LNCS, vol. 5722, pp. 35–44. Springer, Heidelberg (2009)
8. Atzmueller, M., Lemmerich, F.: VIKAMINE - Open-Source Subgroup Discovery, Pattern Mining, and Analytics. Machine Learning and Principles and Practice of Knowledge Discovery in Databases. LNCS, pp. 842–845. Springer, Berlin (2012)
9. Atzmueller, M., Lemmerich, F.: Exploratory pattern mining on social media using geo-references and social tagging information. Int. J. Web Sci. (IJWS), **1/2**(2) (2013)
10. Atzmueller, M., Puppe, F.: Semi-automatic visual subgroup mining using VIKAMINE. Journal of Universal Computer Science **11**(11), 1752–1765 (2005)
11. Atzmüller, M., Puppe, F.: A methodological view on knowledge-intensive subgroup discovery. In: Staab, S., Svátek, V. (eds.) EKAW 2006. LNCS (LNAI), vol. 4248, pp. 318–325. Springer, Heidelberg (2006)
12. Atzmüller, M., Puppe, F.: SD-Map – A fast algorithm for exhaustive subgroup discovery. In: Fürnkranz, J., Scheffer, T., Spiliopoulou, M. (eds.) PKDD 2006. LNCS (LNAI), vol. 4213, pp. 6–17. Springer, Heidelberg (2006)

13. Atzmueller, M., Puppe, F.: A case-based approach for characterization and analysis of subgroup patterns. J. Appl. Intell. **28**(3), 210–221 (2008)
14. Atzmueller, M., Puppe, F., Buscher, H.P.: Exploiting background knowledge for knowledge-intensive subgroup discovery. In: Proceedings of 19th International Joint Conference on Artificial Intelligence (IJCAI-05), pp. 647–652. Edinburgh, Scotland (2005)
15. Becker, M., Mueller, J., Hotho, A., Stumme, G.: A generic platform for ubiquitous and subjective data. In: Proceedings of 1st International Workshop on Pervasive Urban Crowdsensing Architecture and Applications, PUCAA 2013 (2013)
16. Boley, M., Horváth, T., Poigné, A., Wrobel, S.: Listing closed sets of strongly accessible set systems with applications to data mining. Theor. Comput. Sci. **411**(3), 691–700 (2010)
17. Ceci, M., Appice, A., Malerba, D.: Time-slice density estimation for semantic-based tourist destination suggestion. In: Proceedings of ECAI 2010, pp. 1107–1108. IOS Press, Amsterdam, The Netherlands, The Netherlands (2010)
18. Diestel, R.: Graph Theory. Springer, Berlin (2006)
19. Ganter, B., Stumme, G., Wille, R. (eds.): Formal Concept Analysis, Foundations and Applications. Lecture Notes in Computer Science. Springer, Berlin (2005)
20. Han, J., Cheng, H., Xin, D., Yan, X.: Frequent pattern mining: current status and future directions. Data Min. Knowl. Disc. **15**, 55–86 (2007)
21. Han, J., Kamber, M.: Data Mining: Concepts and Techniques, 2nd edn. Morgan Kaufmann, San Francisco (2006)
22. Hotelling, H.: The generalization of student's ratio. Ann. Math. Statist. **2**(3), 360–378 (1931)
23. Jain, A.K., Murty, M.N., Flynn, P.J.: Data clustering: a review. ACM Comput. Surv. **31**(3), 264–323 (1999)
24. Kleinberg, J.: Bursty and hierarchical structure in streams. In: Proceedings of KDD, pp. 91–101. ACM, New York, NY, USA (2002)
25. Klösgen, W.: Advances in knowledge discovery and data mining. In: Fayyad, U.M., Piatetsky-Shapiro, G., Smyth, P., Uthurusamy, R. (eds.) Explora: A Multipattern and Multistrategy Discovery Assistant, pp. 249–271. AAAI, California (1996)
26. Knobbe, A., Fürnkranz, J., Cremilleux, B., Scholz, M.: From local patterns to global models: the lego approach to data mining. In: Proceedings of ECML/PKDD'08 LeGO Workshop (2008)
27. Koperski, K., Han, J., Adhikary, J.: Mining knowledge in geographical data. Commun. ACM **26**, 65–74 (1998)
28. Lakhal, L., Stumme, G.: Efficient mining of association rules based on formal concept analysis. In: Ganter, B., Stumme, G., Wille, R. (eds.) Formal Concept Analysis. LNCS (LNAI), vol. 3626, pp. 180–195. Springer, Heidelberg (2005)
29. van Leeuwen, M., Knobbe, A.J.: Diverse subgroup set discovery. Data Min. Knowl. Discov. **25**(2), 208–242 (2012)
30. Lemmerich, F., Becker, M., Atzmueller, M.: Generic pattern trees for exhaustive exceptional model mining. In: Flach, P.A., De Bie, T., Cristianini, N. (eds.) ECML PKDD 2008, Part II. LNCS, vol. 5212, pp. 277–292. Springer, Heidelberg (2008)
31. Lemmerich, F., Rohlfs, M., Atzmueller, M.: Fast discovery of relevant subgroup patterns. In: Proceedings of 23rd International FLAIRS Conference, pp. 428–433. AAAI Press, Palo Alto, CA, USA (2010)
32. Lindstaedt, S., Pammer, V., Mörzinger, R., Kern, R., Mülner, H., Wagner, C.: Recommending tags for pictures based on text, visual content and user context. In: Proceedings of 3rd International Conference on Internet and Web Applications and Services, pp. 506–511. IEEE Computer Society, Washington, DC, USA (2008)

33. Liu, Z.: A survey on social image mining. Intell. Comput. Inf. Sci. **134**, 662–667 (2011)
34. R Development Core Team: R: A Language and Environment for Statistical Computing. R Foundation for Statistical Computing, Vienna, Austria (2009). http://www.R-project.org
35. Rattenbury, T., Naaman, M.: Methods for extracting place semantics from flickr tags. ACM Trans. Web **3**(1), 1:1–1:30 (2009)
36. Richter, K.-F., Winter, S.: Citizens as database: conscious ubiquity in data collection. In: Pfoser, D., Tao, Y., Mouratidis, K., Nascimento, M.A., Mokbel, M., Shekhar, S., Huang, Y. (eds.) SSTD 2011. LNCS, vol. 6849, pp. 445–448. Springer, Heidelberg (2011)
37. Roitman, H., Raviv, A., Hummel, S., Erera, S., Konopniki, D.: Microcosm: visual discovery, exploration and analysis of social communities. In: Proceedings of IUI, pp. 5–8. ACM, New York, NY, USA (2014)
38. Santini, S., Ostermaier, B., Adelmann, R.: On the use of sensor nodes and mobile phones for the assessment of noise pollution levels in urban environments. In: Proceedings of International Conference on Networked Sensing Systems (INSS), pp. 1–8 (2009)
39. Shneiderman, B.: The eyes have it: a task by data type taxonomy for information visualizations. In: Proceedings of IEEE Symposium on Visual Languages, pp. 336–343. Boulder, Colorado (1996)
40. Sigurbjörnsson, B., van Zwol, R.: Flickr tag recommendation based on collective knowledge. In: Proceeding of the 17th International Conference on World Wide Web, pp. 327–336. WWW '08, ACM, New York, NY, USA (2008)
41. Strehl, A., Ghosh, J., Mooney, R.: Impact of similarity measures on web-page clustering. In: AAAI WS AI for Web Search, pp. 58–64. Austin, TX, USA (2000)
42. Wrobel, S.: An algorithm for multi-relational discovery of subgroups. In: Proceedings of 1st European Symposium on Principles of Data Mining and Knowledge Discovery (PKDD-97), pp. 78–87. Springer, Berlin (1997)
43. Yin, Z., Cao, L., Han, J., Zhai, C., Huang, T.: Geographical topic discovery and comparison. In: WWW 2011, pp. 247–256. ACM, New York, NY, USA (2011)
44. Zhang, H., Korayem, M., You, E., Crandall, D.J.: Beyond co-occurrence: discovering and visualizing tag relationships from geo-spatial and temporal similarities. In: Proceedings of International Conference on Web Search and Data Mining, pp. 33–42. ACM, New York, NY, USA (2012)

Personalised Network Activity Feeds: Finding Needles in the Haystacks

Shlomo Berkovsky[✉] and Jill Freyne

CSIRO, Epping, Australia
{shlomo.berkovsky,jill.freyne}@csiro.au

Abstract. Social networks have evolved over the last decade into an omni-popular phenomenon that revolutionised both the online and offline interactions between people. The volume of user generated content for discovery on social networks is overwhelming and ever growing, and while time spend on social networking sites has increased, the flood of incoming information still greatly exceeds the capacity of information that any one user can deal with. Personalisation of social network activity news feeds is proposed as the solution that highlights and promotes items of a particular interest and relevance, in order to prioritise attention and maximise discovery for the user. In this chapter, we survey and examine the various research approaches for the personalisation of social network news feeds and identify the synergies and challenges faced by research in this space.

1 Introduction

Growth of the Web is relentless and set to continue, even accelerate, as the Web continues to evolve and accommodate new forms of user-generated content [30]. Social networking sites have experienced unprecedented popularity in the past decade and have contributed significantly to increased levels of user generated content that have been fuelling this growth. Social networks, designed to allow anyone to create and distribute content for others to consume, have become rich and diverse sources of information that compete with and complement traditional search engines in the diffusion of information.

Social networks allow users to hook up to streams of information, from trusted individuals, and so act as personal filters for online content. In essence, users hand pick the information sources, whose contributions make up their personal information channel or *news feed*. This methodology, where contributed items or actions are conveniently combined and presented in reverse chronological order worked well, allowing individuals to quickly discover updates and content of interest [5]. The popularity of social networks and the ease of sharing content has, however, swamped the simple news feed aggregation mechanism, as contributions from an increasing numbers of friends and connections flood the feed. The social network structure, which once delivered hand picked content, has become a victim of its own success, much to the frustration of users.

© Springer International Publishing Switzerland 2015
M. Atzmueller et al. (Eds.): MUSE/MSM 2013, LNAI 8940, pp. 21–34, 2015.
DOI: 10.1007/978-3-319-14723-9_2

News feeds of many social networks do not support keyword based search, and users cannot easily find posts relating to topics of their interest. Users can of course unfriend, unfollow, hide, or mute undesired connections in their online social circles who flood their news feed, but the creation of rules and filters requires time and effort and only provides rigid options that turn off the posts from certain users. Add to this potential personal and social unease at unfriending or unfollowing online connections, as the connections are informed of this or this may even become public in some social networks, and we see barriers being raised that practically preclude users from actively curating their own friend list.

Automatic re-organisation of the news feed, aimed at filtering out irrelevant or uninteresting posts and, vice versa, highlighting posts of particular importance and relevance, offers a solid alternative to the manual rules and filters. Thus, more and more researchers[1] in the areas of data mining, machine learning, natural language processing, and social sciences have looked at the social network news feed filtering problem [3]. Several directions were studies under this broad umbrella: what factors make social network posts valuable [19,21], how can the feeds be ranked in a generic manner [12,18,27], and what semantic Web approaches can alleviate the ranking task [6]. However, many of these works stumbled upon a major obstacle – posts that are interesting for one user are not interesting for another user. That is, importance and relevance are user dependent, such that the feeds need to be filtered in a *personalised* manner.

Confirmation of the need for personalisation was found by Alonso et al. in a study that solicited opinions on how interesting or uninteresting a set of tweets from reputable organisations (BBC, New York Time, Reuters, and so forth) [1]. The authors asked five raters to indicate interestingness of more than 2,000 tweets and, unsurprisingly, the overall inter-rater agreement was close to 0. That is, the perceived interestingness of tweets was found to be subjective and user-dependent. To learn more, Deuker and Albers investigated user priorities for news feed content in a series of structured interviews aimed at uncovering the factors that determine content attractiveness in social networks [11]. They concluded that content attractiveness stems from two major factors: recipient's *interest* in the topic of a content item and in the *author* (or poster) of the item. Clearly, both factors are user-dependent – a user may be interested in certain topics more than in others and appreciate content posted by some users more than by others. The design space of news feed personalisation was studied by Chen et al. in [8]. The authors focussed on Twitter and, in line with the findings of [11], outlined three considerations which were shown to correspond to user interest or satisfaction. The *source* of the tweets and the recipient's *topics of interest* were highly relevant; in addition, the *popularity* of the tweets was also a driving factor.

---

[1] Notwithstanding, the under-water part of this iceberg includes applied research done by large-scale social networks, such as Facebook, LinkedIn, and Twitter. They put much effort into feed filtering and develop proprietary solutions, but these are most often not disclosed due to the commercial sensitivity and competitiveness.

So, what information can facilitate relevance judgements and scoring needed for personalisation? Let us name a few. A post or news feed item typically has an *author*, some *textual content*, and has often experienced *social approval* judgements, such as 'likes', 'shares', 'retweeets', or 'replies'. Posts also often contain some additional *content*, e.g., hashtags, URLs, pointers to other users, and so forth. Moving away from individual posts, social networks, by their very nature, have an underlying *network structure* that reflects explicit friendships, follower/followee relationships, or other articulated connections established on the network. Finally, there is a moderate amount of *interaction* data that, if captured and mined properly, can provide a rich source of information for user model creation. For example, content viewing can imply user interest, whereas viewing user profiles of others and direct communication with others can provide implicit indicators to supplement the explicit connection information. Finally, in social network user profiles often include unverifiable *background information* related to user demographics, location, preferences, skills, interests, and many other facets, which can be leveraged to inform the feed personalisation. Aggregating all these into a robust and accurate personalisation mechanism is not a straightforward task.

In this chapter we survey and bring together the state-of-the-art approaches for news feed personalisation. Our survey uncovers three dominant themes, under which the research fits. The first theme focuses primarily on those users who contribute posts and content, those who potentially see their posts, and the links between the two. User-to-user tie strength research examines how two users have interacted in the past to determine if one user's posts should be given high priority in the news feed of the other. The second theme deals with the actual content of the posts and will sound familiar for readers knowledgeable about content based data mining and information retrieval, where the content of items (typically, text included in or linked to by a post) is examined to determine correlation to user interests or a query. The third theme details a set of works that look at the graph and structure of the underlying social network of both users and their posts, to determine similarity and relevance as criteria for inclusion of posts in the news feed. We also briefly touch upon other considerations, such as temporal information, use of latent factor models, and use of mobile apps and devices. Upon surveying the state-of-the-art works, we synthesis them, highlight popular motives coming through, raise emerging topics and research questions, and outline promising directions for future research.

2   Feed Personalisation – The Current State of Play

Let us start with some formalisation of the feed personalisation problem. Feed personalisation can be naturally considered as either a top-K recommendation or a re-ranking problem. Let us denote by N the set of candidate items that can potentially be included in the feed, e.g., all the activities carried out on the social network by the user's friends or all the tweets posted/retweeted by the user's followees. With no personalisation applied to the feed, these are typically shown in a reverse chronological order. Personalisation implies selecting a subset $K \in N$, such that $|K| \ll |N|$, which correspond to items of a higher importance for the recipient of the the feed. Given this formalisation of the personalisation problem, the recommendation task entails scoring the $|N|$ candidate items and selecting $|K|$ top-scoring items on behalf of the user. Alternatively, the re-ranking task entails re-ordering the $|N|$ chronologically ordered candidates, i.e., scoring all the candidate items, keeping the $|K|$ top-scoring items on top of the list, and removing the remaining $N \backslash K$ items. The distinction between the recommendation and re-ranking solutions is quite blurry, such that in the rest of this chapter we will consider works that belongs to both without explicitly classifying them.

The central role of accurate news feed item scoring mechanism in the personalisation process is evident. In the following subsections, we survey several approaches for feed item scoring. We first elaborate on the user-to-user relationships; then, we examine works that incorporate text and content factors; then, social network and graph representation related considerations; and, finally, we introduce additional factors such as temporal information and constraints posed by the mobile device used by a feed recipient.

2.1   User-to-User Relationships

One of the pivotal considerations in identifying and scoring items of relevance look at the relationships between the user who performed the action or posted the content, and the recipient of the feed. Several works looked into the quantification of the online *tie-strength* between two social network users.

The trail-blazing work in this area was performed by Gilbert and Karaholios [15]. They proposed seven dimensions, that represented the strength of the relationship, or tie-strength, between pairs of Facebook users: *intensity* - amount of communication exchanged between the two; *intimacy* - use of intimacy and familiarity language in the communication; *duration* - period of time since the two established the online ties; *reciprocal* - resources, apps, and information shared between the two; *structural* - common groups and networks, or shared interests; *emotional* - gifts or congratulations exchanged between the two; and *distance* - similarity of religion, education, or political views. Using these dimensions, the authors derived 70 features and populated them using the observable online communication between the two users. The overall tie-strength score was computed as a linear combinations of these features. The tie-strength model was trained and the weights of individual features were determined using more than 2,000 explicit judgements provided by 35 participants (questions like "how strong is your relationship with X?"). It was found that the intimacy dimension accounted for more than 30 % of the tie-strength score, whereas the most highly correlating individual features mirrored the duration of relationship: days since first and last communication. An offline study achieved predictive error smaller than 10 %, showing the validity of the developed model.

A similar vein of research, aimed at predicting professional and personal closeness of an enterprise social network users, was done by Wu et al. [34]. They derived 60 features predicting user-to-user closeness and split these into five categories:

subject user - activity of the user who performed the action; *target user* - activity of the recipient of the feed; *direct interaction*: intensity of direct interaction between the subject and the target user; *indirect interaction* - intensity of indirect interaction between the two through common friends; and *corporate* - distance between the two in the organizational structure. Also here the overall closeness score was computed as a linear combination of the individual features. For the model training, the authors collected more than 4,000 explicit professional and personal closeness scores ("how closely are you currently working with X?" and "how likely are you to talk with X about your non-work life?". Interestingly, the most important group of features was found to be the direct interaction between the two users: it accounted for close to 40 % of weight for the professional and 48 % of weight for the personal closeness. Predictive accuracy was evaluated, and the error was 18 % for the professional and 22 % for the personal closeness.

Another perspective on user-to-user relationship was taken by Jacovi et al. in [20]. The authors focussed on the interest in a user, i.e., curiosity in the activities done by that user. The interest reflects a directional asymmetric relationship, which may differ from closeness and tie-strength. They proposed four implicit indicators that may signal interest in a user: directly *following* the user, *tagging* the user in a people-tagging service, *viewing content* contributed by the user, and *commenting* on the user's posts. Close to 120 participants were presented with lists of their online acquaintances and asked to select users of interest. Out of the above four indicators, tagging users was the most strongly correlated with interest, followed by direct following, and then by viewing accessing and commenting that were comparable. It should be noted that the observed correlation between the two explicit signals (tagging and following) and the interest level was almost double the correlation of the two implicit signals (viewing content and commenting).

A model that combines the features of [15] with the interaction-based weighting of [34] was proposed by Berkovsky et al. in [5]. In addition to the tie-strength score, the model also incorporated user preferences towards certain social network actions, such as posting/viewing content, commenting on posts of others, uploading images, and so forth. The underlying social network was an experimental portal of people engaged in a healthy living program. As the portal was fully controlled, the authors trained the model against the observed feed clicks as implicit interest indicators, and conducted an online inter-group study with a subset of the participants being exposed to personalised feeds. More than 500 feeds with clicks were reconstructed and analysed, and it was found that personalising the feeds increases user interactions, extends the duration of portal sessions, and boosts the contribution of user-generated content.

2.2   Text and Content Factors

Predicting the tie-strength of the user who posted a content item or performed a network activity is only one facet of the overall importance of the item in the news feed [11]. Other things that should be taken into consideration include the content of the posted item. The term 'content' embraces both the immediate text included in the posts and other information, such as URLs, pointers to other users, tags, and more.

Paek et al. collected Facebook data pertaining to the perceived importance of individual news feed entries ("how important do you feel this feed item is for you?") [24]. 24 users provided close to 5,000 explicit feed item importance judgements, which were discretised into binary important/unimportant labels. Using the observable Facebook logs, the authors mined and populated 50 predictive features across three groups: *social media* - metadata, number of comments, views, and likes, inclusion of URLs, and temporal information about posts and users; *text* - processed content of the post and previous communication between the two users, such as n-grams and $tf \times idf$ vectors; and *background information* - static information about the users' location, education, activities, interests, mined from their public Facebook profiles. An SVM classifier was trained and an ensemble model of all the features achieved close to 70 % accuracy, which dropped to 63 % when textual features were removed from the predictive model. This drop highlights the importance of the textual content of the posts/activities in the feed scoring model.

A similar model for ranking of tweets on Twitter was developed and evaluated by Uysal and Croft in [32]. They aimed specifically at the tweet ranking task and derived a suite of features that were split into four categories: *poster* - reputation, popularity, and activity of the person who posted the tweet; *content* - inclusion of hashtags, URLs, user mentions, and other emotional signals in the tweet; *text* - novelty and language model of the tweet content; and *recipient* - relation and past interactions between the recipient and the poster of the tweet. These categories of features were used individually as well as in combination, and evaluated offline using a corpus of more than 2,500 previously observed retweets. The best classification accuracy was achieved by the combined model (F-measure of 0.72). Content features were the top-performing category (F-measure of 0.5), while the performance of the pure textual features was surprisingly poor (F-measure of 0.04), perhaps due to the noisy nature of the text included in tweets. The dominance of the content features was re-affirmed in a tweet ranking accuracy evaluation.

Shen et al. proposed a method for a personalised interest-based reordering of tweets of a user's followees [29]. User interests were determined by analysing the tweets published and consumed by the user, and modelling the topics of these tweets. The reordering incorporated five feature models: *temporal* - freshness of the tweet; *influence* - authority of the poster: number of poster's followers and followees, number of lists on which the poster appears, and age and verification of the poster's account; *quality* - length, URL, and hashtags of the tweet, as well as the number of retweets; *match* - match of the tweet to the interests of the recipient; and *social* - number of retweets and replies between the poster and the recipient. An ensemble model incorporating the above features was built and trained, considering the tweets that were retweeted or replied as interesting and aiming to prioritise and position these at the top of the tweet list. The reordering model was found to outperform the non-personalised and time-based models with respect to several evaluation metrics. Interestingly, the most important features were the freshness of the tweet, the number of retweets, and the number of poster's followers.

Rather than sorting and filtering tweets, they can be grouped into lists, each bearing a degree of relevance to various topics of interest of the target user. This method was studied by Burgess et al. in [7]. The set of topics and users posting tweets related to these topics was extracted by analysing the established follower-followee links, clustering users in dense sub-graphs, and mining the textual content of tweets posted by these users. The method was evaluated against manually created lists containing several hundreds of users. The automatically extracted lists of users and topics resembled the manual ones, with the observed F1 scores hovering between 0.7 and 0.8. The authors also developed several heuristics for assessing the relevance of individual tweets to the extracted lists. One heuristic was underpinned solely by the textual content of the tweets (unigrams and $tf \times idf$ vectors), whereas the other considered also the included hashtags. The heuristics demonstrated a comparable degree of accuracy, with a slight preference toward those based on the textual content. The heuristics were also shown to be robust to noise, such that their accuracy only degraded slightly when the level of noise was as high as 50 %.

2.3   Network and Graph Structure Factors

The value of textual features in microblogs is significantly lower than in typical social media, as the characteristics such as the length of posts, presence of acronyms, and high dynamicity of topics make text analytics difficult. Chen et al. in [9] proposed that the extracted features needed to be augmented with information mirroring the structure of the network. The authors devised a personalised tweet ranking model, based on the observable retweets as implicit interest indicators. The model encapsulated a suite of features categorised into four groups: *relation* - friendship between the poster and recepient, overlapping of their followees, and number of mentions in previous tweets; *content relevance* - relevance of the tweet content to the recepients' status, retweet, or hashtag histories; *content* - length, hashtags, and URLs of the tweet; and *poster authority* - number of poster's followers, followees, mentions, and status updates. These features were fed into a latent factor model that was evaluated using a corpus of more than 100,000 retweets. It was found that the model was consistently superior to several baselines and achieved average precision of 0.76. Also, the combined model substantially outperformed the individual models undeprinned by single groups of features.

The work of Feng and Wang used the graph-based model of Twitter to rank tweets [13]. The nodes of the graph encapsulated the users (both the tweet poster and the recipient) and the tweets themselves, whereas the edges expressed the poster-recipient and recipient-tweet relationships. Additional features about the tweets (hashtags, URLs, age, popularity), users (similarity, mentions, reputation, probability to retweet and be retweeted), as well as user-tweet relationships (user profile vs tweet content similarity, mentions, hashtags) were mined. These features were used to train a factorisation model, aimed at predicting the retweet probability for a given author, recipient, and tweet. The model was deployed to rank tweets according to their predicted retweeting probability and was evaluated against a corpus of more than 2.1 million retweets done by more than 28,000 users. The average precision of the combined model was around 42 %, whereas a comparison of individual features showed that the author-recipient and recipient-tweet edges dominated the user and tweet nodes. This large-scale evaluation highlights the value encapsulated in the Twitter graph structure.

In [35], Yan et al. proposed a graph-theoretic model for personalised tweet recommendations. The recommender leverages a heterogeneous graph model consisting of a graph of users and a graph of tweets. In both sub-graphs, the nodes represent the users and the tweets, respectively, while the edges reflect the degree of their similarity. The user-to-user similarity is established based on the commonality of their followees, while the similarity of tweets is computed using their semantic content. Additionally, edges that connect the user and tweet sub-graphs indicate the original poster and the retweeters of a tweet. The nodes of the two sub-graphs are initially scored using the personalised PageRank algorithm, and then co-ranked, such that tweet score correspond to the scores of its poster and retweeters and, vice versa, user score correspond to the scores of the tweets they posted and retweeted. The model was applied for the tweet ranking task and evaluated using a corpus of more than 55 million retweets. The results demonstrated good ranking (nDCG greater than 0.5 for various sizes of the list) and classification (precision close to 62 %) accuracy and outperformed several personalised ranking competitors.

2.4   Other Considerations

The challenge of ranking social updates on LinkedIn using click stream data was studied by Hong et al. in [17]. The authors evaluated three families of predictive models. The family of *linear* models included a feature-based model (features of the source user, recipient, and the update were used), a bias model (source user, recipient, and update category bias considered), and a temporal model. The family of *latent factor* models included a matrix and a tensor factorisation models, as well as their variants capitalising on a suite of manually crafted user features, such as seniority, connectedness, frequency and recency of visits, and so on. Lastly, since the above two families are optimised against different loss functions, they were combined using a *pairwise learning* model. The models were evaluated offline using LinkedIn's interaction logs. Out of the linear models, the bias model achieved the highest precision, 0.53–0.60, for different training/testing splits. Tensor factorisation model with features was the top-performing latent model, with precision at the range of 0.59–0.65. Pairwise learning managed to combine the strengths of the two models and demonstrated precision scores hovering between 0.62 and 0.66.

Another clue to the importance of feed items lies in the temporal information, as user interests may drift over time. Two temporal dependencies – in performing social network activities and estimating user-to-user relevance – were studied by Freyne et al. in [14]. The authors exploited an offline dataset of user interactions with the news feed of an enterprise social network and evaluated several short-term, long-term, and combined temporal models. Although the combined model demonstrated the best performance, it was found that short-term models were predictive of user-to-user relevance, while long-term models were found suitable for assessing the relevance of network actions for users. In other words, implicit user-to-user tie-strength score were found more volatile than the observed behaviour and interaction patterns of users.

The use cases of social networks are becoming increasingly mobile, such that the specific question of personalising the feed on a mobile device becomes relevant. This question was addressed by Cui and Honkala in [10]. They developed several content-based approaches (collaborative approaches cannot run on the client, as no access to information of others is available) that score feed entries and predict future clicking probabilities according to the items clicks observed in the past. A personalised PageRank predictor, a Bayesian predictor, and an ensemble model were evaluated in a 4-week live user study involving 40 participants. The Bayesian predictor was found to outperform the PageRank predictor individually, while the highest accuracy was achieved by the ensemble model. Furthermore, it was found that incorporating the time dimension in the model substantially improved the accuracy of the obtained results.

3 Discussion and Emerging Topics

Having surveyed a range of works on personalisation of social network activity feeds, we would like to summarise them in a concise manner. Table 1 presents the key contributions grouped by their underlying social environment, the predictive features that were used by the personalisation mechanism, the data against which the mechanism was trained, and the evaluation metrics used.

As can be seen from the table, most feed personalisation work published to date had conducted their evaluation on public social networks and only several used proprietary networks or ad-hoc communities established for the evaluation purposes. We would like to highlight the prevalence of Twitter as the chosen evaluation platform. We posit that this is attributed to two reasons. Firstly, the sheer volume of tweets faced by Twitter users means that Twitter is the "poster boy" for feed personalisation. Initial works in personalisation on Twitter were for the followee recommendation functionality and ways of expanding your network, but this has quickly been followed by works that support filtering through personalisation. Twitter is an attractive platform on which to carry out evaluations also due to the availability of data and API for easy crawling [22,23]. We note also the strong dominance of implicitly provided training data, e.g., feed clicks, retweets, and replies, over explicitly labelled data. Indeed, it is unreasonable to expect users to explicitly annotate their network activity feed items, unless this is rewarded or directly related to their mainstream social network interactions.

Considering the groups of used features, we highlight the large number of works leveraging the network structure in the prediction process. This is not surprising, given that relationships and links established on a social network inherently reflect user interest in other users and/or in the content they contribute [16]. Of direct relevance to this is the reputation or authority of content

Table 1. Summary of feed personalisation works. Predictive features include user *activity* (actions of the poster and recipient), observed *interactions* (messages, tagging, replies), *network* structure (relationships between users, common friends), *graph* features (PageRank score, node connectedness), *textual* content of the posts, other *content* (hashtags, URLs, statuses), *reputation* (number of followees, followers, mentions), *temporal* information (posting frequency, account age), *similarity* (between the post and user interests, between the interests of users), and *static* information (location, gender, interests). Evaluation metrics include click-through rate (CTR), classification accuracy (CA), prediction accuracy (PA), and ranking accuracy (RA) metrics, usability questionnaire, computation time, and contributed content [28].

Work	Network	Features	Training	Metrics
Berkovsky et al. [5]	Health-related community	Action, activity, interaction	Feed clicks	CTR, CA, RA, contribution
Burgess et al. [7]	Twitter	Network, text	Explicit lists	CA, PA, novelty
Chen et al. [9]	Twitter	Text, network, content, similarity, reputation	Retweets	CA, CTR
Cui and Honkala [10]	Ad-hoc user community	Graph, interaction, temporal	Feed clicks	Usability, PA
Feng and Wang [13]	Twitter	Text, content, network, reputation, activity, similarity, temporal	Retweets	CA, time
Freyne et al. [14]	Enterprise social network	Activity, temporal, interaction	Feed clicks	CTR, RA
Hong et al. [17]	LinkedIn	Reputation, temporal, graph, network, static	Feed clicks	CA
Paek et al. [24]	Facebook	Text, network, static, content	Explicit judgements	CA, PA
Shen et al. [29]	Twitter	Temporal, reputation, content, similarity, activity, network	Retweets, replies	RA, CA
Uysal and Croft [32]	Twitter	Reputation, activity, network, interaction, content	Retweets	CA, RA
Yan et al. [35]	Twitter	Graph, network, text	Retweets	RA, CA

posters, which serves as an indicator of their salience on the network and is exploited in a number of works. It is important to note that the network structure can be pre-computed ahead of time, with relationship scores determined by previously observed interactions, and a simple lookup will determine an item's expected relevance. Then, we note the use of content features – both the textual content and other content, such as hashtags and URLs. The content may serve as a direct predictor for whether the post or tweet will be relevant to the interests of the recipient. There is often a computational requirement involved in this process, where content, not included in a post needs to be fetched and examined before a relevancy judgement can be made. In similar to various predictive tasks, temporal information is important and, as expected, these features are exploited in a number of approaches.

Two groups of features, increasingly leveraged in recent works, should be discussed. The first refers to features extracted from a graph-based representation of the social network in hand. These features supplement network features and encapsulate graph-based metrics derived from representing the data as a graph [31]. The potential of these features has been shown in other domains, such that their use for filtering and ranking of network feeds is natural and timely. Also in this case, the social network graph is often known and can be pre-computed and referenced to achieve personalisation in real time. In addition, the graph allows for content discovery, as relationships with users who are non-direct friends, but to whom the user is close in a network graph, can be seen to provide relevant feed items for consumption, facilitating serendipitous people and content discovery.

The second group of features deals with user-to-user interactions. Although some insight can be obtained from the established network links, a fine-grained quantification of user-to-user relationships should be derived from their mutual interactions, e.g., viewing the contributed content, mentioning each other, sending direct messages or retweeting, or even interacting with the same group of users [33]. We conjecture that features reflecting observable network interactions will gain an increasing popularity. Indeed the work of Berkovsky et al. [5] showed how relationships between different pairs of users can vary. Users may have friends whose photos they like to see, or whose blog articles they like to read. Being able to extrapolate strength and context of relationships between users is a valuable means for not only filtering out users but filtering out posts, in order to satisfy the differences identified in what users find interesting.

The vast majority of the work on feed personalisation use classification accuracy metrics. These highlight the requirement to simply predict whether a post will be of interest for the recipient or not, rather than determining the exact level of interest. Understanding the performance of an algorithm in a classification task is most suitable when the system aims to filter items from the news feed, rather than making explicit recommendations for items to consume. The second most-frequently used metric is ranking accuracy. This is natural, considering that the size of the feed shown to users is typically limited, and the importance of correctly ranking items in the feed is paramount [28]. If the exact level of interest in an item can be predicted, short lists of recommended items can further reduce the effort required to discover the most interesting items. Metrics incorporating click-through-rates or predicting the scores of feed items are also used, but less popular than classification and ranking metrics.

We are seeing increased diversity in the formats and media shared on social networks, and yet the research that we are seeing is primarily limited to a few commercial products. While news feeds are considered very natural and established features of widely popular social networks such as Facebook and Twitter, there is a gap in personalisation for social networks such as Pinterest, Instagram or Flickr, on which photo sharing is the aim but network, but hashtags and user tie-strengths could also be applied with ease. In addition, we need to prepare for the arrival of new types of networks and media and more work is needed to uncover effective ways to merge the news feeds from multiple sources and networks to simplify things even further. Some progress into the aggregation of news feeds across multiple social network has been made by Summify[2] and several other commercial sites which link network accounts, monitor the target user's social graph, and email the digest of few most relevant stories per day.

Much of the work that was overviewed in our survey focuses on the mechanics and technicalities of using data mining and machine learning to identify interesting posts. Little work was uncovered that discussed the user interface needs associated with personalised news feeds. The metrics mentioned above lend themselves to different interface types, as discussed. Users have become increasingly familiar with the chronological lists that include all posts, interesting and boring together. Thus, there is a great need for work into novel interfaces, visualisations, and control mechanisms (consider tag clouds, multimedia plugins, and approaches recently proposed by Twingly[3] and Fidgt[4]) desired by users so that the feed personalisation is recognised as the time saving, productivity tool that it is originally aimed to be.

A number of years ago Facebook started to filter their news feeds, by removing content from users that it deemed the individual was less interested in. They faced an incredible push-back from the users, who were far from impressed by the lack of control and rigidity of the new filtering features. In essence, the users felt that this filtering contributed to the so-called *filter bubble* [25], a situation where the network decides on behalf of the users what feed items they are interested in, such that the users become isolated from cultural or ideological bubbles different to their opinions and viewpoints. This situation is unacceptable from ethical perspective and the social networks should find the gold spot between personalising the feed and limiting information exploration at the same time. Likewise, social feeds may pose a privacy threat, as they expose potentially sensitive information about activities in the user's close social circles that can accessed by untrusted parties or used inappropriately. Hence, privacy considerations should also be taken into account when filtering news feed items [4].

All in all, we feel that the research into social network activity feed personalisation is relatively in its infancy. Several solid algorithmic techniques were developed and thoroughly evaluated so far. Having said that, social network designers should keep in mind that their networks are user-facing systems. As such, much attention should be devoted to user aspects of personalisation: what do users find interesting and valuable [19]; how should the feeds be visualised and presented [26]; how do they prefer to interact with the feed [33]; does the feed answer their needs in the most encompassing and unobtrusive way [2]. We conjecture that these topics will have an increased exposure in the coming years and encourage researchers to consider these questions in their work.

---

[2] http://www.summify.com.

[3] http://www.twingly.com/screensaver.

[4] http://sourceforge.net/projects/fidgtvisual/.

References

1. Alonso, O., Marshall, C.C., Najork, M.: Are some tweets more interesting than others? #hardquestion. In: HCIR (2013)
2. Bao, J., Mokbel, M.F., Chow, C.: Geofeed: a location aware news feed system. In: ICDE, pp. 54–65 (2012)
3. Berkovsky, S.: Network activity feed: finding needles in a haystack. In: MSM, p. 1 (2013)
4. Berkovsky, S., Borisov, N., Eytani, Y., Kuflik, T., Ricci, F.: Examining users' attitude towards privacy preserving collaborative filtering. In: DM.UM (2007)
5. Berkovsky, S., Freyne, J., Smith, G.: Personalized network updates: increasing social interactions and contributions in social networks. In: Masthoff, J., Mobasher, B., Desmarais, M.C., Nkambou, R. (eds.) UMAP 2012. LNCS, vol. 7379, pp. 1–13. Springer, Heidelberg (2012)
6. Bontcheva, K., Rout, D.P.: Making sense of social media streams through semantics: a survey. Semant. Web 5(5), 373–403 (2014)
7. Burgess, M., Mazzia, A., Adar, E., Cafarella, M.J.: Leveraging noisy lists for social feed ranking. In: ICWSM (2013)
8. Chen, J., Nairn, R., Nelson, L., Bernstein, M.S., Chi, E.H.: Short and tweet: experiments on recommending content from information streams. In: CHI, pp. 1185–1194 (2010)
9. Chen, K., Chen, T., Zheng, G., Jin, O., Yao, E., Yu, Y.: Collaborative personalized tweet recommendation. In: SIGIR, pp. 661–670 (2012)
10. Cui, Y., Honkala, M.: A novel mobile device user interface with integrated social networking services. Int. J. Hum.-Comput. Stud. 71(9), 919–932 (2013)
11. Deuker, A., Albers, A.: Who cares? content sharing on social networking sites: a grounded theory study. In: PACIS, p. 156 (2012)
12. Duan, Y., Jiang, L., Qin, T., Zhou, M., Shum, H.: An empirical study on learning to rank of tweets. In: COLING (2010)
13. Feng, W., Wang, J.: Retweet or not?: personalized tweet re-ranking. In: WSDM, pp. 577–586 (2013)
14. Freyne, J., Berkovsky, S., Daly, E.M., Geyer, W.: Social networking feeds: recommending items of interest. In: RecSys, pp. 277–280 (2010)
15. Gilbert, E., Karahalios, K.: Predicting tie strength with social media. In: CHI, pp. 211–220 (2009)
16. Guy, I., Zwerdling, N., Ronen, I., Carmel, D., Uziel, E.: Social media recommendation based on people and tags. In: SIGIR, pp. 194–201 (2010)
17. Hong, L., Bekkerman, R., Adler, J., Davison, B.D.: Learning to rank social update streams. In: SIGIR, pp. 651–660 (2012)
18. Huang, H., Zubiaga, A., Ji, H., Deng, H., Wang, D., Le, H.K., Abdelzaher, T.F., Han, J., Leung, A., Hancock, J.P., Voss, C.R.: Tweet ranking based on heterogeneous networks. In: COLING, pp. 1239–1256 (2012)
19. Hurlock, J., Wilson, M.L.: Searching twitter: separating the tweet from the chaff. In: ICWSM (2011)
20. Jacovi, M., Guy, I., Ronen, I., Perer, A., Uziel, E., Maslenko, M.: Digital traces of interest: deriving interest relationships from social media interactions. In: ECSCW, pp. 21–40 (2011)
21. Lage, R., Denoyer, L., Gallinari, P., Dolog, P.: Choosing which message to publish on social networks: a contextual bandit approach. In: ASONAM, pp. 620–627 (2013)

22. McCormick, T.H., Lee, H., Cesare, N., Shojaie, A.: Using twitter for demographic and social science research: tools for data collection. Sociological Methods and Research (2013)
23. Morstatter, F., Pfeffer, J., Liu, H., Carley, K.M.: Is the sample good enough? comparing data from twitter's streaming api with twitter's firehose. In: ICWSM (2013)
24. Paek, T., Gamon, M., Counts, S., Chickering, D.M., Dhesi, A.: Predicting the importance of newsfeed posts and social network friends. In: AAAI (2010)
25. Pariser, E.: The Filter Bubble: What the Internet is Hiding from You. Penguin UK, London (2011)
26. Parra, D., Brusilovsky, P., Trattner, C.: See what you want to see: visual user-driven approach for hybrid recommendation. In: IUI, pp. 235–240 (2014)
27. Sarma, A.D., Sarma, A.D., Gollapudi, S., Panigrahy, R.: Ranking mechanisms in twitter-like forums. In: WSDM, pp. 21–30 (2010)
28. Shani, G., Gunawardana, A.: Evaluating recommendation systems. In: Ricci, F., Rokach, L., Shapira, B., Kantor, P.B. (eds.) Recommender Systems Handbook, pp. 257–297. Springer, New York (2011)
29. Shen, K., Wu, J., Zhang, Y., Han, Y., Yang, X., Song, L., Gu, X.: Reorder user's tweets. ACM TIST 4(1), 6 (2013)
30. Susarla, A., Oh, J., Tan, Y.: Social networks and the diffusion of user-generated content: evidence from youtube. Inf. Syst. Res. 23(1), 23–41 (2012)
31. Tiroshi, A., Berkovsky, S., Kâafar, M.A., Vallet, D., Chen, T., Kuflik, T.: Improving business rating predictions using graph based features. In: IUI, pp. 17–26 (2014)
32. Uysal, I., Croft, W.B.: User oriented tweet ranking: a filtering approach to microblogs. In: CIKM, pp. 2261–2264 (2011)
33. Wilson, C., Boe, B., Sala, A., Puttaswamy, K.P.N., Zhao, B.Y.: User interactions in social networks and their implications. In: EuroSys, pp. 205–218 (2009)
34. Wu, A., DiMicco, J.M., Millen, D.R.: Detecting professional versus personal closeness using an enterprise social network site. In: CHI, pp. 1955–1964 (2010)
35. Yan, R., Lapata, M., Li, X.: Tweet recommendation with graph co-ranking. In: ACL, vol. 1, pp. 516–525 (2012)

Ontology-Enabled Access Control and Privacy Recommendations

Marcel Heupel[1]([✉]), Lars Fischer[2], Mohamed Bourimi[3], and Simon Scerri[4]

[1] Department Business Information Systems IV, University of Regensburg,
Regensburg, Germany
Marcel.Heupel@ur.de
[2] Research Group IT Security Management, University of Siegen, Siegen, Germany
fischer@wiwi.uni-siegen.de
[3] MT AG, Ratingen, Germany
Mohamed.Bourimi@mt-ag.com
[4] Organized Knowledge Group, Fraunhofer IAIS, Bonn, Germany
scerri@iai.uni-bonn.de

Abstract. Recent trends in ubiquitous computing target to provide user-controlled servers, providing a single point of access for managing different personal data in different Online Social Networks (OSNs), i.e. profile data and resources from various social interaction services (e.g., LinkedIn, Facebook, etc.). Ideally, personal data should remain independent of the environment, e.g., in order to support flexible migration to new landscapes. Such information interoperability can be achieved by ontology-based information representation and management. In this paper we present achievements and experiences of the di.me project, with respect to access control and privacy preservation in such systems. Special focus is put on privacy issues related to linkability and unwanted information disclosure. These issues could arise for instance when collecting and integrating information of different social contacts and their live streams (e.g., activity status, live posts, etc.). Our approach provides privacy recommendations by leveraging (1) the detection of semantic equivalence between contacts as portrayed in online profiles and (2) NLP techniques for analysing shared live streams. The final results after 3 years are presented and the portability to other environments is shortly discussed.

Keywords: Personal Information Management · Decentralised social networking · Access control · Privacy recommendations · Pervasive computing · di.me · Linkability

1 Introduction

Online Social Networks (OSN) became the dominating web-service of the early 21st century. Storing and controlling personal data of a large degree of the online population they became both, places to improve social interaction, and dangerously large single points of failure waiting for the next big data incident

© Springer International Publishing Switzerland 2015
M. Atzmueller et al. (Eds.): MUSE/MSM 2013, LNAI 8940, pp. 35–54, 2015.
DOI: 10.1007/978-3-319-14723-9_3

to happen. OSN help with organising social life, not the least of them being the contact list that reminds of imminent birthdays. Additionally, the big central OSN, envision not only to provide organisational services, but a complete portfolio of communication channels.

The current state of social networks though seems disappointing in many ways, especially in terms of interaction between different providers, which is not sufficiently supported. As a result, the digital world is divided into different social spheres and individuals have to manually connect their different social representations. However, recently there is a recognizable trend moving towards the unification, integration and digitisation of personal information from various sources. Many social media platforms are starting to allow the users to import/export personal data or sharing profiles and activities amongst different platforms. For instance, users are able to share their status messages in LinkedIn directly with their Twitter channel, and so on. The above provides two clear advantages. On one hand, users profit in terms of easier management and sharing of related information across their personal and social spheres from one online service to another (e.g., by easily importing their own profile and contact information, personal data, etc.). On the other hand, platform providers also become able to optimise their added-value services by analysing collected data with the explicit consensus of respective users, given that they accept the underlying privacy policy in-order to use such facilities.

Concerning the management of multiple identities in different networks, and thereby keeping such identities strictly separate, is a complicated task. We have currently a growing list of examples, where even people without having full access to a social networks database, can infer quite personal information, ranging from sexual orientation to pregnancy status. As it has been shown in particular for users on Flickr and Twitter, accounts can be re-identified with an error rate of just 12 % across distinct social networks [1]. Furthermore, linkability issues can also arise when retrieving the users' profiles, even when anonymisation solutions are applied (e.g., by using pseudonyms, removing identifying data, etc.) as shown in [2]. Consequently, advanced mechanisms for the management of multiple identities in different networks are necessary.

An example for a sophisticated approach, fostering the management of multiple identities and integration of information from various OSNs, is the European research project digital.me[1]. One of the central outcomes of the project is the di.me system [3], a decentralised application for privacy-enhanced social networking. The system is based on a semantic model and supports users in managing information and identities in Online Social Networks (OSNs).

Derived from experiences gained during the runtime of the digital.me project, we will present requirements, challenges and appropriate solutions for the building of similar systems, especially focusing on semantic access control and how privacy issues related to linkability and unwanted information disclosure (i.e., information being shared in personal live streams) can be avoided. The core objective of our approach is to trigger privacy warnings/recommendations to

---

[1] http://www.dime-project.eu/.

the user in potentially risky situations. In order to detect such situations, information from online profiles and personal live streams is being analysed using Natural Language Processing (NLP) techniques, e.g., to detect if contacts are using multiple profiles.

In the sections hereafter, our contribution is structured as follows: The project di.me as an exemplary system that meets a large subset of the requirements is detailed in its concepts and techniques in Sect. 2, together with a detailed discussion of the specific requirements addressed in the context of this work. A specific approach, developed in the context of the di.me project is presented in Sect. 3. Section 5 compares the results to the state-of-the-art before Sect. 6 concludes the paper and gives some ideas about future plans.

2 Problem and Requirements Analysis

2.1 di.me Background Information

The European research project digital.me was concerned with the research on privacy-enhancing technologies when using online services. It targeted to enhance users privacy by helping users keeping track of digital footprints scattered over different networks. During the project lifetime of three years a social networking platform has been developed as the core result of cooperative research by different partners from academic and industrial fields across Europe. The platform implements many sophisticated privacy-enhancing concepts, from anonymous communication support to user guidance in the User Interface (UI). As one of the major outcomes of the project was the *di.me userware*, has also been published as open source[2]. Main objective of the di.me project was to develop an advanced platform for user-controlled and privacy-ensuring communication and data management, which had to be considered in the whole design and development process on different levels. The project leveraged semantic technologies and open data exchange for personal identity and information management and decentralized communication networks. di.me contributed to the creation of a more user-friendly environment for user-controlled, decentralised, and smarter networking and information exchange for different communities. This also fostered adoption of similar solutions by professionals from outside the ICT sector.

2.2 Semantic Access Control Engine

Ideally, the design of access control for private data should be flexible and independent of the target host. Personal data should also remain independent of environmental specifics, e.g., in order to support flexible migration to new landscapes. di.me's Access Control Engine which incorporated a two-layered access control in order to decouple the semantic core from the hosting environment. It thereby ensures that personal data and the associated ontology-based access rights remain flexibly decoupled from the underlying environment (Fig. 1).

---

[2] https://github.com/dime-project/meta.

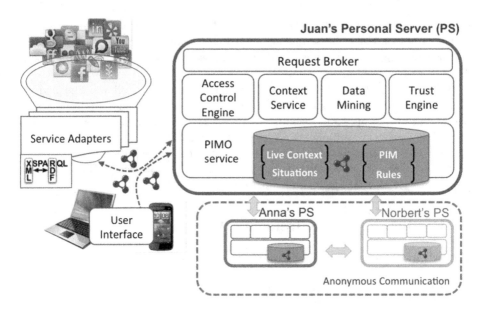

Fig. 1. di.me high level architecture

As we describe in [4], the Access Control Engine in the overall system architecture is considering relationships to other components, especially the Trust Engine. A developed trust metric is used to support context-aware access control decisions as we describe in [5]. Two main points of usage were thereby identified: First, it helps the user assessing access rights to his/her data, by the provision of privacy recommendations. Then, after a while of usage of the di.me userware, when the Trust Engine learned enough about the users privacy preferences[3], the user is able to define special rules to automatically share specific data to trusted contacts. Both of this access rules are defined with the help of the PPO, the Privacy Preference Ontology (cf. Subsect. 3.1).

The Access Control Engine has been integrated in all interaction flows connected to the disclosure of information. In the di.me architecture, a request broker forwards incoming calls either to the Trust Engine or to the Access Control Engine. In both cases, the Access Control Engine is involved since the Trust Engine involves it indirectly. The business logic is than carried out by involving the di.me controllers layer as well as the storage layer (for persisting roles, permissions, roles and further access control data incl. those managed with the help of the ontology model). The opposite direction is also supported which means that access control decisions could include the Trust Engine.

In summary, the objectives of the Access Control Engine as declared in [4] target to model trust, privacy and security throughout the system. Thereby it is an important tool in order to e.g., perform access control decisions, enforce

---

[3] The system calculates trust values based on the privacy of exchanged content and its distribution, or can be set manually. For more information see e.g. [6] and [7].

Table 1. Requirement categories for privacy-enhancing OSN platforms

(R1)	Integrated Personal Information Management
(R2)	Secure, Privacy-respecting Sharing of Personal Information
(R3)	Intelligent User Support with Context Sensitive Recommendations and Trust Advisory
(R4)	Transparent Multi-Platform UI, and
(R5)	Integration of Existing Services

secure authentication and authorization, enable "User Controlled Data Sharing", define (trust enhanced) privacy settings and consider privacy policies. However, in the context of this work we will specifically concentrate on the specification of the ontologies in order to support semantic access control and privacy support, and on means to enhance users privacy with recommendations and advisory in cases of potential linkability issues and unwanted information disclosure.

2.3 Requirements for Avoiding Linkability and Unwanted Information Disclosure

The research nature of the di.me project implies to ensure innovation at research level, so various existing solutions have been compared against an analytically elaborated list of user requirements. The results of the requirements-driven approach based on the comparison of existing decentralised solutions[4] has been summarized in [8]. With involvement of end-users[5] and companies[6] we gathered five requirement categories R1–R5 (see also Table 1), whose focus is to support the integration of the personal information, by providing security and privacy-enhancing mechanisms. Further, intelligent user support with context sensitive recommendations and trust advisory [5] will be provided. Trust metrics guide the user to avoid risky behaviour when disclosing private information. Anonymous data disclosure, data withdrawal and policies foster privacy and trust [2] and approaches for secure end-user driven deployment in public clouds were elaborated [9]. The personal data is modelled using a comprehensive set of integrated and multi-domain ontologies [3].

The current reference architecture fulfils the identified requirement categories R1–R5 and implements a semantic core with data mining, semantic mapping and reasoning, supporting an intelligent management of personal data and communication history including recommendations how to take advantage of the personal sphere. Intelligent user interfaces on desktop and mobile devices promoting the

---

[4] As in cases of many alpha releases it is almost impossible to get reliable results for an evaluation, we restricted the set of systems for this analysis to the following five stable system implementations: Diaspora, Friendica, Jappix, Kune and StatusNet.

[5] The project implemented a user-driven design process and requirements were refined several studies and validations.

[6] Cf. http://www.dime-project.eu for details on the consortium application partners.

intuitive usage of powerful semantic and privacy-technologies enable the user to monitor, control, and interpret personal data.

With respect to the requirement category R2 (Privacy Respecting Sharing of Personal Information), by considering leveraging advances in the implementation of the semantic core (category R1, Integrated Personal Information Management) and integration of existing services (category R5), further potential enhancements were identified. A promising[7] enhancement of R1, R2 and R5 is based on experiences from various projects requiring solving the classical problem described by Krontiris and Freiling who mention in [10] that *"As argued by Spiekermann and Cranor [...], privacy by policy offers the minimum degree of protection and systems utilising such solutions need to make users aware of privacy risks and offer them choices to exercise control over their personal information"*.

Even though di.me is a decentralised solution, linkability problems and unwanted information disclosure could be not fully avoided (at least due to potential user's naivety). An important functionality of the di.me userware is the support of digital faces or partial identities. This means one user could be in contact with the same person using various pseudonyms at the same time as shown in Fig. 2. When maintaining multiple digital identities, one of the most difficult tasks is to make linking those identities as difficult as possible (e.g., linking of a pseudonymous account to a real name account could compromise the users privacy). Therefore, the di.me userware needs to provide support when sharing information via different identities to uphold unlinkability. In order to be able to give such support a first requirement is to identify contacts that might be the same person, or contacts that might know each other and could collaborate. Besides the vulnerability on application level, another issue emerging thereby is that these faces or identities link to the personal server (PS) of the respective user.

Besides the provision of privacy recommendations when sharing profile information and files, it is crucial to consider one of the most important communication mechanisms in current OSNs is the so called "microblog", or "status message", in di.me referred to with the term LivePost. Those post are spontaneous typed short text messages about current activity or location (e.g. "checkin") and can thereby easily contain more private information than intended, about the user him-/herself, but also about others. Consequently, one of the privacy enhancing mechanisms targeted in di.me is to analyse such LivePosts, identify private information and give meaningful privacy recommendations, in order to prevent unintended information disclosure and to and to provide awareness for such privacy threats.

3   Semantic Access Control and Privacy Advice

In di.me, ontologies and information extraction techniques are used by the semantic core to provide various privacy-related features. Besides regulating

---

[7] Also in terms of innovation!.

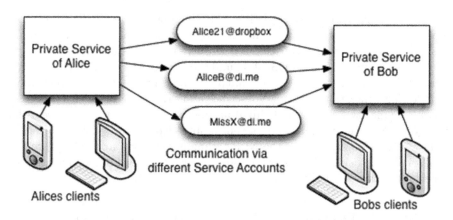

Fig. 2. Multiple Service-Account communication scheme potentially leading to linkability

access rights, ontologies are also used in combination with personal information extracted by computational techniques to raise personal and social awareness and help respect the users' privacy. The enabled privacy-related features include:

- Ontology-based authorisation gives the user better control and helps keeping track which data has been shared, can be shared, under which circumstance, and with whom.
- the retention of privacy control outside a person's userware so as to warn against the resharing of privacy-sensitive information in between di.me users.
- extending the targeted integration of heterogeneous online profiles to potentially identify people behind anonymous/pseudonymous identities.
- analysing online microposts to determine whether the type of information being shared might breach the user's or any other implicated person's privacy.
- determining whether information that the user shares online in anonymous profiles and microposts can reveal their identity so as to prevent unintended linkability, based on the semantics of concerned profile attributes and posted content types.

These capabilities are integrated in the di.me Trust & Privacy Engine which, based on discovered knowledge aggregated in the semantic core, is responsible for triggering privacy warnings and recommendations to the user in the UI.

To enable the above-listed features, we were required to: introduce access control, privacy and trust awareness at the semantic representation layer (i.e. extend the di.me ontologies); provide a component that is able to compare local and online profiles with the aim of identifying semantic-equivalence (i.e. identify multiple pseudonymous profiles for a unique person); and employ Natural Language Processing (NLP) techniques to extract context- and activity-related information from a user's microposts.

The first subsection below describes the di.me ontologies necessary to provide the privacy-related features. The ontologies are integrated within the di.me ontology framework, which consists of a set of re-used, extended as well as new vocabularies provided by the OSCA Foundation (OSCAF)[8]. The technique for the identification of known people behind anonymous profiles is then discussed in the following subsection, where we describes a metric that computes a weighted semantic similarity based on individual profile attributes, in order to reveal the identity of anonymous people and prevent unintended linkability for the user. In the last subsection, we discuss the NLP techniques that are applied on streams of personal and social online posts, in order to detect the unintended sharing of potentially sensitive information.

3.1   Semantic Technology for Privacy Control and Awareness

Ontology-Based Access Rights. Different information in the di.me integrated personal information sphere is subject to different access rights, as defined and controlled by the user. An ontology-based approach to a flexible authorisation system enables the user's sharing preferences to be stored separately to the information resources they want to provide, or restrict, access to. The primary purpose of the Privacy Preference Ontology (PPO) [11] is to enable Web users to create fine-grained privacy preferences for their public online data. The Web takes an "open-world" view to personal information whereby, unless restricted, all information is assumed to be accessible by everyone. In di.me, the PPO is adopted with a "closed-world" view, i. e., all information is initially inaccessible unless the owner explicitly grants access rights.

Figure 3 shows how vocabulary from the following ontologies is combined with the PPO to achieve the required functionality. The Personal Information Model Ontology[9] (PIMO) enables the representation of a person's entire information cloud, including people they know and the different groups they can assign them to. The Sharing Ontology[10] (NSO) models information sharing irrespective on the platform on which it is shared, through the definition of an access space and a number of agents that can access it. The Annotation Ontology[11] (NAO) represents high-level, generic annotation, including privacy levels for information items and trust levels for agents. The Account Ontology[12] (DAO) represents the concept of a personal account, which can represent, receive or act on behalf of a person online as a special form of an agent. To cover the requirements, vocabulary from these ontologies is used in combination as follows. As a special instance of *pimo:Agent*, the user can define multiple privacy preferences (*nso:hasPrivacyPreference*), each of which identifies one or more informational resources (*ppo:appliesToResource*) that can be shared through an Access

---

[8] http://www.oscaf.org/.
[9] http://www.semanticdesktop.org/ontologies/2007/11/01/pimo/.
[10] http://www.semanticdesktop.org/ontologies/2009/11/08/nso/.
[11] http://www.semanticdesktop.org/ontologies/2007/08/15/nao/.
[12] http://www.semanticdesktop.org/ontologies/2011/10/05/dao/.

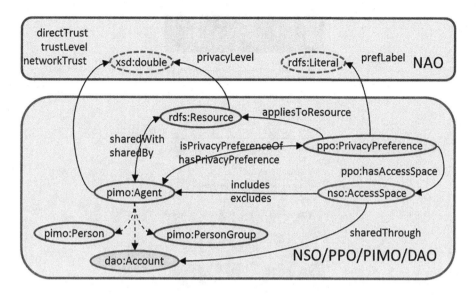

Fig. 3. Access rights, privacy and trust in digital.me

Space(*ppo:AccessSpace*). The latter refers to a number of other agents and is populated through white-lists by (*nso:includes*) and black-lists (*nso:excludes*) in the di.me user interface. In di.me, PPO privacy preferences can take the form of *Databoxes* and *ProfileCard*. The former refers to different subsets of information elements, the latter to different subsets of personal profile attributes, which can be shared with specific groups of agents. For usability purposes, each privacy preference can be assigned a label (e.g. 'ProjectFiles' Databox, or a 'Business' ProfileCard).

Besides information authorisation, ontologies have also been employed to make the user more aware of what information has been shared and with whom. The NSO ontology (*nso:sharedWith*) is also used to mark which resources have been shared with other agents, regardless of the current privacy preferences. Whereas the latter define with whom resources can be shared, the former keep a record of with whom they have already been shared. An inverse property (*nso:sharedBy*) keeps track of the provenance of a resource. The property *nso: sharedThrough* was introduced in line with di.me's multiple identity paradigm - information items can be shared using one of possibly many personal accounts (and associated identities/ProfileCards) through the same di.me userware. Ontologies are also used to assign trust and privacy levels to agents and resources respectively. The NAO ontology[13] provides vocabulary for registering both user-defined (*nao:directTrust*) and network-derived (*nao:networkTrust*) trust levels, which are then combined in socio-personal trust levels(*nao:trustLevel*). Different privacy levels can also be assigned to resources using *nao:privacyLevel*.

---

[13] http://www.semanticdesktop.org/ontologies/nao/.

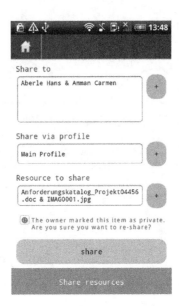

Fig. 4. Warning against re-sharing private item with third parties

Combined, privacy and trust levels can identify many sensitive information-sharing events in which the user could be warned.

The use of metadata to store privacy and trust-related information introduces a possibility to extend their application outside of the user's personal information sphere.di.me introduced the use of ontology-based "sticky policies", whereby the original owner is still able to retain some form of control after sharing items with other di.me users. Thus, when a user attempts to re-share an item whose resharing was not intended by the original owner, the new owners can be warned that they might be committing a privacy breach. To enable this functionality, item privacy levels are shared as accompanying metadata, together with core attributes such as creation/modification date and creator. For items that are assigned a privacy level that is more than moderate (0.6 or larger, where 1 signifies very private, 0 public), if their recipient attempts their re-sharing (with third parties) by any di.me-controlled means, they will be warned of a potential privacy issue (Fig. 4).

Discovering and Linking Multiple Identities. An appropriate semantic equivalence metric is one of the requirements for any aspiring self-integrating system [12]. As with other aspects of personal information, once two or more items (in this case, personal identities managed as profiles) are determined to be equivalent (i.e., they represent the same person), they are semantically 'reconciled' through a single representation at the level of the PIMO Ontology (introduced in the previous section), which maintains a representation of the user's Personal Information Model (PIM).

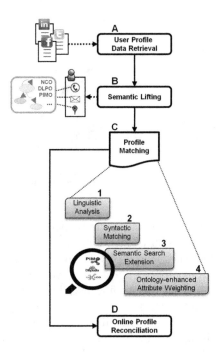

Fig. 5. Semantic equivalence approach overview

The entire approach for the elicitation, transformation, comparison, discovery and integration of equivalent user profiles is depicted in Fig. 5. The first stage (A) simply retrieves structured and semi-/unstructured profile information (belonging to both the users and their contacts) from multiple online accounts, through the provided APIs. This information is lifted (stage B) onto the di.me ontologies, i.e., each retrieved profile results in a single instance of the Contact Ontology (NCO)[14], based on the *nco:PersonContact* concept.

In the third stage (C), newly retrieved profiles are matched against the existing instances. The use of ontologies and the underlying Resource Description Framework (RDF)[15] as a data representation format means that the matching we pursue considers both syntactic as well as semantic similarities between two profiles. We now provide a short summary of the 4-step (C1-4) matching process [13]:

- Step C1 performs linguistic analysis on profile attributes that contain complex unstructured information (e.g., a person's description or physical address), in order to extract further knowledge from non-atomic attribute values (e.g., region, city and country from a person's full physical address; first name and surname from a person's full name).

---

[14] http://www.semanticdesktop.org/ontologies/2007/03/22/nco/.
[15] http://www.w3.org/RDF/.

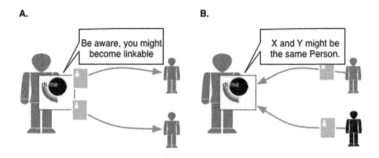

Fig. 6. Use of semantic matching of persons in di.me

- Step C2 performs syntactic matching through the use of a string matching algorithm, specifically a variant of the Monge and Elkan recursive field matching algorithm [14]. This metric is used for comparing textual attributes (and extracted named entities) from a newly-retrieved user profile against existing NCO person profiles, in order to find matches (e.g. same name, surname, username, address, country, etc.).
- Step C3 extends the search semantically, by comparing the retrieved named entities to representations in existing repositories, with the aim of (i) finding direct semantic matches so as to be in a better position to compare profile attributes and (ii) finding indirect semantic matches, i.e., semantically-related albeit not equivalent profile attributes (e.g. location 'North Rhine-Westphalia' vs. location 'Bonn'). The main repository over which this search is performed is the PIM, being the user's own personal knowledge base, and thus a source of personally-relevant entities. If the PIM search does not return any matches, the search extends to public datasets that form part of the Linked Open Data (LOD) cloud[16], such as DBPedia[17].
- Step C4 consists of a weighted metric that combines the results of both syntactic and semantic matching. The matched attributes are given an ontology-based weight, depending on their nature (e.g. profile attributes marked as inverse functional, such as personal email addresses, have a higher weight than other attributes, such as country of residence) and on the target domain of the online account(e.g. a professional user profile e.g. LinkedIn, is more likely to match other profiles in skills-oriented social networks).

The above-described matching process (stage C) enables the di.me userware to determine, with an 82 % precision rate [13], whether a profile (belonging to either the user or to one of their contacts) is semantically equivalent to a person that is already known in their PIM. Thus, a positive match will result in the newly-retrieved profile to be reconciled (stage D) with profile 'occurrences' through a unifying *pimo:Person instance* (refer to Fig. 3).

---

[16] http://lod-cloud.net/.
[17] http://wiki.dbpedia.org/OnlineAccess.

Fig. 7. Warning showing up when a potential privace breach is detected

Through the above technique, the di.me userware is able to automatically identify matching person profiles. This also includes cases when one or more of the profiles is intended to be anonymous (e.g. a person uses a secondary identity), based on a number of matching personal details (e.g. same date-of-birth, city, affiliation or a combination of which). The identification of profiles/identities that potentially belong to a same person has two use-cases. In the first (shown as A in Fig. 6), the user is warned that a secondary/anonymous identity can easily be linked to another of their identities, perhaps unintentionally, based on a number of overlapping attributes. The second use-case (shown as B in Fig. 6), shows how in privacy-sensitive sharing scenarios (e.g. the user is about to share information with untrusted people) the system will warn that one intended recipient could be someone they are already acquainted with.

Detecting Privacy-Sensitive Online Posts. The analysis of streams of online posts in order to gain information about personal interests and activities, so as to provide meaningful suggestions to the user, is an often-discussed topic in the literature [15]. One of the main goals of di.me is to raise the user's awareness with respect to their auto-disclosure of privacy-sensitive information online, through sharing activities such as microblogging. Users are posting short snippets of information about what they are currently doing, where and with whom. Apart from some functions which can be more easily controlled, like checking in a specific place, tagging particular people in a post, etc., a majority of such posts consists solely of unstructured, free text. This practice can easily result in compromising the privacy of another person, e.g., posting a status that claims that you are at the beach with Anna, when Anna should be at the workplace. In di.me we try to prevent accidental and harmful information disclosure, whether it affects the user posting or people implied in the online post, by considering both structured posts (e.g. check-ins, tags) and free text.

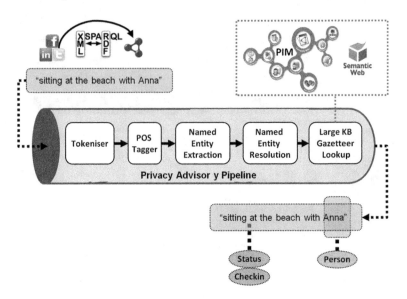

Fig. 8. The text analysis pipeline used by the Privacy Advisor

The di.me Information Extraction (IE) component employs Natural Language Processing (NLP) techniques to parse textual microposts before they are shared, and decompose them into different types of (possibly co-occurring) posts, based on the Live Post Ontology (DLPO)[18]. For example, the text in the example above, coupled with a picture showing John with Anna at the beach, is decomposed into: a *dlpo:ImagePost* referring to the actual picture, a *dlpo:CheckIn* referring to the geo-location of the beach, and a *dlpo:Status* containing the textual message. In addition, once matched to the corresponding PIM item, person Anna is marked as a related resource through *dlpo:relatedResource*. Being machine-processable representations, DLPO instances are then processed by di.me's privacy advisory system to detect sensitive information disclosure and provide suitable warnings. In the example shown in Fig. 7, before sharing a status message on both di.me and Facebook (exported through the API), the user is warned that they might want to reconsider publishing information related to a third party's location and activity. Figure 8 shows the IE pipeline used by the privacy advisory component in di.me. Micropost processing starts with Named Entity Extraction (NEE), upon which, syntactic matching techniques similar to the ones described in the previous section are employed for Named Entity Resolution (NER), i.e. to discover which of these entities (including people) are already known in the PIM[19]. Core concepts of the PIMO ontology are very similar to the generic entities typically extracted by NEE algorithms

---

[18] http://www.semanticdesktop.org/ontologies/2011/10/05/dlpo/.

[19] Although an authoritative scientific evaluation is still forthcoming, preliminary experiments suggest the following indicative F-measures: Persons: 51 %, Locations: 60 %, Events: 45 %.

(e.g. people, organisations, locations), but also include more personal (or group) entities (e.g. projects, events, tasks). NEE taggers based on gazetteers are a good fit for entity extraction where a personal Knowledge Base (KB) (e.g. the PIM), may feedback the algorithm with new entities created directly by the user, or from entities resolved and retrieved from an external KB.

4   Validation Results

During the projects lifetime several validations of different scales have been carried out. Validations and interviews of smaller scale have been frequently performed, in order to get an early impression of the user acceptance of implemented features. With respect to the focus of this paper, the last two validation phases are of major interest, which were addressing the general public and CRM customers over the time of 3 months. The first phase, being conducted from July till October 2013, was addressing the general acceptance, usability and utility of the userware. A total of 278 user tested the system, while only 127 responded to the questionnaire. The mean age of the testers was 32 years and the distribution male:female was roughly 2:1. Around 64 % of the respondents worked in an IT-related field and about 70 % of the respondents claimed to have profound knowledge in the are of social media. A remarkable result of the validation was, that 81 % appreciated the idea of privacy recommendations, while 65 % indicated that they would be interested in such a product.

The second larger validation was especially concerned with the trust and privacy advisory. It has been conducted in August 2013 and had a total of 447 testing users, while only 23 filled out the questionnaire. The validation was mainly focusing on the three acceptance conditions "Utility, Usability and Privacy concerns" of the trust and privacy advice when exchanging sensitive documents. The evaluation identified a high demand of trust-enabling technologies in the CRM sector. The strongest positive response received the privacy condition, while the results also identifies that further efforts should increase the usability and investigate a better understandable presentation of the concepts.

5   Comparison to Related Work

Social platforms need some degree of information disclosure in order to achieve the sharing goals and to sustain social engagement, as Palen and Dourish state in [16]. However, it could also compromise the users' privacy and thereby sometimes have irreversible fatal consequences. Server-centric approaches imply that the server is the central point of information exchange, allowing service providers to create fully-fledged user-profiles. With respect to information disclosure, decentralised social networks offer more control for the end-user [17]. Many research projects put their focus on the development of concepts and solutions for bypassing the problem of profile building, for centralised as well as decentralised systems. Nevertheless, approaches developed for decentralised platforms could be also be applied on centralised topologies. In this respect, the approach presented in this paper is still valid for being ported to server-centric environments.

Table 2. General approaches

	Celino	Steiner	Cano	Choudhury	Chang	Abel	Zoltan	Passant	Liu	Ritter	di.me
Keyword Extraction	○	○	●	●	○	○	●	○	○	○	●
Topic Extraction	○	○	●	○	●	●	●	●	○	●	●
NEE (Events)	○	◐	○	●	◐	●	○	○	○	◐	●
NEE (People)	○	◐	●	●	●	●	●	○	●	●	●
NEE (Activities)	○	◐	○	○	◐	○	○	○	○	○	●
NEE (Locations)	◐	◐	●	○	●	●	●	●	●	●	●
NER	○	○	○	○	○	○	●	○	○	○	●

Although di.me compares to a number of other decentralised solutions (refer to requirements analysis presented in [8]), the combination of techniques as proposed by the approach presented in this paper has, to the best of our knowledge, not been rivalled in contemporary literature. A wider overview of social sharing and collaboration platforms can be found in [18]. There, Smith et al. also propose an approach for decentralised and privacy-preserving micro-sharing that use Web standards as ontologies. However, di.me's added value lies in the provision of a privacy advisory that applies natural language processing techniques to extract 'dynamic' user context- and activity-related information from online posts, by matching them to items in a personal semantic repository. This information supplements the more 'static' privacy-aware information that is defined by the user and their trusted network, including item privacy levels and person trust levels. By reasoning over the combined static and dynamic ontological knowledge, di.me can thus provide context-aware privacy warnings and recommendations. Several social-based recommendation techniques also employ computational techniques to extract information from social data [19], e.g. to recommend movies, products or to collect opinions [20–22]. Nevertheless, none of the cited efforts consider leveraging the extracted information to provide privacy recommendations.

Whereas the IE technique applied by di.me over socially-shared data is comparable to previous efforts in the area, it targets the extraction of more items that may reveal privacy-sensitive information about the current activities of the user [15]. Table 2 outlines a comparison against a number of other IE techniques performed on informal, 'noisy' micropost communication: A [23], B [24], C [25], D [26], E [27], F [28], G [29], H [30], I [31], J [32]. Although some form of linguistic analysis (keyword/topic extraction, NEE for various entity types) is performed by all, none target all the potential activity-revealing information identified in the first column of the table. In particular, apart from di.me, only Chang and Sun [27] target the four major named entities targeted for extraction by di.me, albeit partially. NER is only performed by Zoltan and Johann [29]. In addition, only the latter cited links extracted named entities to existing semantic representations, namely to concepts from within the LOD Cloud. Apart from linking extracted entities to open community knowledge, in di.me we also exploit the availability of personal information in the PIM. This has the obvious advantage that it consists only of personal data, making it easier to determine equivalence between entities mentioned in the microposts and existing PIM items.

6   Conclusions and Future Work

In this paper, we addressed the enhancement of privacy in OSNs by aggregating contacts' information and their live streams from various social online services in order to provide valuable privacy recommendations. This was concretely discussed for the decentralised platform developed in the EU FP7 project di.me. The semantic core and the trust engine, which are central components of the di.me userware, have been enhanced by adding intelligent information extraction techniques and utilised in order to unintended information disclosure, possibly compromising the privacy of the user or his/her contacts.

A deep analysis of the state-of-the-art that should further identify innovation potential in the di.me research project, as well as the involvement of users and companies was the basis for the requirement gathering.

In our approach we presented three mayor contributions. The first, was the definition of ontology based access rights, followed by the semantic equivalence detection, which is being used, on the one hand, to detect contacts possibly using multiple different accounts and to trigger suggestions to combine all information to an integrated profile representation. On the other hand, the same mechanism helps the user to maintain unlinkability of own identities, if it is applied to respective profiles of the user him/herself. The third presented contribution was the provision of privacy recommendations when disclosing information through microblogging and related communication channels. Therefore advanced NLP mechanisms are being utilised in order to process live text inputs and to detect sensitive information before posting to other services. With those contributions, the concepts developed in the context of the di.me project offer a clear improvement over related work by leveraging semantic capabilities and NLP techniques in order to provide advanced privacy recommendations.

The reference implementation of the di.me userware is able to retrieve profile data from different platforms (e.g. LinkedIn, Facebook, Twitter and Google+). The main challenge is to discover if two or more online descriptions of a person are semantically equivalent by combining several required techniques and also the improvement of NLP-based recognition capabilities. Most popular techniques for semantic equivalence detection are of a syntactic nature, i.e. comparing string values of various profile attributes. The approach presented in this paper ensures more accurate results by an extension of the matching capabilities to also compare attributes 'semantically' by using clearly-specified meanings of profile attribute types, as well as through an exploration of their semantic (in addition to syntactic) relatedness. In order to detect semantically equivalent representations of a person the users' data is integrated at the Personal Information Model (PIM) level. The PIM is an abstraction of the possibly multiple occurrences of the same data as available on multiple online accounts and devices. So it represents one singular digital identity containing unique personal data such as their, files, task lists and emails, amongst others.

Ongoing and future work focuses on merging NLP with the semantic equivalence capabilities in order to support privacy recommendation/advisory in various scenarios. For example, the user can be warned that they are: (i) disclosing sensitive information, such as password, accounts, visa number, private telephone number etc., (ii) providing contradicting information over different channels (e.g. current location), which can lead to loss of credibility and trust, and (iii) getting a contact request from a person who, based on different aggregated information in chats, live streams etc., appears to be very close to or possibly even the same person as a known (untrusted) contact.

These warnings will be presented to the user through intelligent interfaces for various platforms. di.me already provides the possibility for people to personally define rules that trigger privacy warnings/suggestions. The mechanism and user interface for defining and checking these rules in real-time is also driven by ontologies and the semantic representations of the PIM [33]. This is perhaps the biggest advantage of the semantic approach behind di.me; few other knowledge representation models offer the sophistication required for this task. Privacy rules can fire not only when the system determines that an online post may compromise someone's privacy, but also when the context-recognition component determines that information that is publicly being shared by the user (e.g. shared documents, personal contact information) should be withheld due to privacy-compromising circumstances (e.g. untrusted person in the area). The ontology-driven functionalities, including the privacy rules in action, have been successfully demonstrated in [34], and can also be viewed online[20].

Further future extensions will also consider live context data (e.g., position, nearby people, activities, time) and data gathered by analysing what a person's contacts have been posting. Di.me can thereby give even more valuable recommendations and also warn about contradictory situations, e.g., when Anna's post about being sick at home 30 min ago is followed by the micropost shown earlier, whereby a friend implicates her presence at the beach. Finally, additional work will also target the non-intrusive presentation of the respective recommendations and advisory.

References

1. Narayanan, A., Shmatikov, V.: De-anonymizing social networks. In: Proceedings of the 30th IEEE Symposium on Security and Privacy (S&P 2009), pp. 173–187. IEEE Computer Society, Oakland, 17–20 May 2009
2. Bourimi, M., Heupel, M., Westermann, B., Kesdogan, D., Planaguma, M., Gimenez, R., Karatas, F., Schwarte, P.: Towards transparent anonymity for user-controlled servers supporting collaborative scenarios. In: Ninth International Conference on Information Technology: New Generations (ITNG), pp. 102–108, April 2012
3. Scerri, S., Gimenez, R., Herman, F., Bourimi, M., Thiel, S.: digital.me - towards an integrated personal information sphere. In: Federated Social Web 2011 (2011)
4. Bourimi, M., Scerri, S., Planaguma, M.: A two-level approach to ontology-based access control in pervasive personal servers. Research report (2011)

---

[20] http://vimeo.com/dimeproject/videos.

5. Heupel, M., Fischer, L., Bourimi, M., Kesdoğan, D., Scerri, S., Hermann, F., Gimenez, R.: Context-aware, trust-based access control for the digital.me userware. In: 5th International Conference on New Technologies, Mobility and Security (NTMS), pp. 1–6 (2012)
6. Heupel, M., Bourimi, M., Kesdoğan, D.: Trust and privacy in the di.me userware. In: Kurosu, M. (ed.) HCII/HCI 2013, Part III. LNCS, vol. 8006, pp. 39–48. Springer, Heidelberg (2013)
7. Heupel, M., Bourimi, M., Kesdogan, D.: The di.me trust approach for supporting collaborative scenarios. In: 1st Workshop on Security in highly connected IT systems (DEXA 2014 workshops), pp. 1–5. IEEE CSP, June 2014
8. Thiel, S., Bourimi, M., Giménez, R., Scerri, S., Schuller, A., Valla, M., Wrobel, S., Frà, C., Herman, F.: A requirements-driven approach towards decentralized social networks. In: Proceedings of the International Workshop on Social Computing, Network, and Services (2012)
9. Karatas, F., Bourimi, M., Barth, T., Kesdogan, D., Gimenez, R., Schwittek, W., Planaguma, M.: Towards secure and at-runtime tailorable customer-driven public cloud deployment. In: 2012 IEEE International Conference on Pervasive Computing and Communications Workshops (PERCOM Workshops), pp. 124–130, March 2012
10. Krontiris, I., Freiling, F.: Integrating people-centric sensing with social networks: a privacy research agenda. In: 2010 8th IEEE International Conference on Pervasive Computing and Communications Workshops (PERCOM Workshops), pp. 620–623, 29 March–2 April 2010
11. Sacco, O., Passant, A.: A privacy preference ontology (PPO) for linked data. In: Linked Data on the Web Workshop at 20th International World Wide Web Conference. ACM Press (2011)
12. Ray, S.R.: Interoperability standards in the semantic web. J. Comput. Inf. Sci. Eng. ASME **2**, 65–69 (2002)
13. Cortis, K., Scerri, S., Rivera, I., Handschuh, S.: An ontology-based technique for online profile resolution. In: Jatowt, A., Lim, E.-P., Ding, Y., Miura, A., Tezuka, T., Dias, G., Tanaka, K., Flanagin, A., Dai, B.T. (eds.) SocInfo 2013. LNCS, vol. 8238, pp. 284–298. Springer, Heidelberg (2013)
14. Monge, A., Elkan, C.: The field matching problem: algorithms and applications. In: Proceedings of Second International Conference on Knowledge Discovery and Data Mining, pp. 267–270 (1996)
15. Scerri, S., Cortis, K., Rivera, I., Handschuh, S.: Knowledge discovery in distributed social web sharing activities. In: Proceedings of the 2nd Workshop on Making Sense of Microposts (MSM2012), WWW 2012 (2012)
16. Palen, L., Dourish, P.: Unpacking "privacy" for a networked world. In: CHI '03: Proceedings of the SIGCHI Conference on Human Factors in Computing Systems, pp. 129–136. ACM Press, New York (2003)
17. Bourimi, M., Ossowski, J., Abou-Tair, D.D., Berlik, S., Abu-Saymeh, D.: Towards usable client-centric privacy advisory for mobile collaborative applications based on BDDs. In: 4th International Conference on New Technologies, Mobility and Security (NTMS) Conference, pp. 1–6 (2011)
18. Kleek, M.V., Smith, D.A., Shadbolt, N., Schraefel, M.C.: A decentralized architecture for consolidating personal information ecosystems: the webbox. In: PIM 2012, January 2012. Event Dates: 11 Feb 2012
19. King, I., Lyu, M.R., Ma, H.: Introduction to social recommendation. In: Proceedings of the 19th International Conference on World Wide Web, WWW '10, pp. 1355–1356. ACM, New York (2010)

20. Shen, E., Lieberman, H., Lam, F.: What am i gonna wear? Scenario-oriented recommendation. In: Proceedings of the 12th International Conference on Intelligent User Interfaces, IUI '07, pp. 365–368. ACM, New York (2007)

21. Wartena, C., Slakhorst, W., Wibbels, M.: Selecting keywords for content based recommendation. In: Proceedings of the 19th ACM International Conference on Information and Knowledge Management, CIKM '10, pp. 1533–1536. ACM, New York (2010)

22. Papangelis, A., Galatas, G., Makedon, F.: A recommender system for assistive environments. In: Proceedings of the 4th International Conference on PErvasive Technologies Related to Assistive Environments, PETRA '11, pp. 6:1–6:4. ACM, New York (2011)

23. Celino, I., Dell'Aglio, D., Valle, E.D., Huang, Y., Lee, T., Park, S., Tresp, V.: Making sense of location based micro-posts using stream reasoning. In: Proceedings of the 1st Workshop on Making Sense of Microposts (#MSM2011) at ESWC, pp. 13–18 (2011)

24. Steiner, T., Brousseau, A., Troncy, R.: A tweet consumers' look at twitter trends. In: Proceedings of the 1st Workshop on Making Sense of Microposts (#MSM2011) at ESWC (2011)

25. Cano, A.E., Tucker, S., Ciravegna, F.: Capturing entity-based semantics emerging from personal awareness streams. In: Proceedings of the 1st Workshop on Making Sense of Microposts (#MSM2011) at ESWC, pp. 33–44 (2011)

26. Choudhury, S., Breslin, J.: Extracting semantic entities and events from sports tweets. In: Proceedings of the 1st Workshop on Making Sense of Microposts (#MSM2011) at ESWC, pp. 22–32 (2011)

27. Chang, J., Sun, E.: Location: How users share and respond to location-based data on social networking sites. In: Proceedings of the AAAI Conference on Weblogs and Social Media, pp. 74–80 (2011)

28. Abel, F., Gao, Q., Houben, G.-J., Tao, K.: Analyzing user modeling on twitter for personalized news recommendations. In: Konstan, J.A., Conejo, R., Marzo, J.L., Oliver, N. (eds.) UMAP 2011. LNCS, vol. 6787, pp. 1–12. Springer, Heidelberg (2011)

29. Zoltan, K., Johann, S.: Semantic analysis of microposts for efficient people to people interactions. In: Proceedings of the 10th Roedunet International Conference (RoEduNet), pp. 1–4 (2011)

30. Passant, A., Bojars, U., Breslin, J., Hastrup, T., Stankovic, M., Laublet, P.: An overview of smob 2: open, semantic and distributed microblogging, pp. 303–306 (2010)

31. Liu, X., Zhang, S., Wei, F., Zhou, M.: Recognizing named entities in tweets. In: Proceedings of the 49th Annual Meeting of the Association for Computational Linguistics: Human Language Technologies, HLT '11, vol. 1, pp. 359–367. Association for Computational Linguistics, Stroudsburg (2011)

32. Ritter, A., Clark, S., Mausam, Etzioni, O.: Named entity recognition in tweets: an experimental study. In: EMNLP (2011)

33. Scerri, S., Debattista, J., Attard, J., Rivera, I.: A semantic infrastructure for personalisable context-aware environments. AI Mag. Forthcoming **35** (2014)

34. Scerri, S., Rivera, I., Debattista, J., Thiel, S., Cortis, K., Attard, J., Knecht, C., Schuller, A., Hermann, F.: di.me: A context-aware information system (demo). In: Proceedings of the 11th European Semantic Web Conference on The Semantic Web: Research and Applications, ESWC'14 (2014)

Refining Frequency-Based Tag Reuse Predictions by Means of Time and Semantic Context

Dominik Kowald[1,2](✉), Simone Kopeinik[2], Paul Seitlinger[2], Tobias Ley[3], Dietrich Albert[2], and Christoph Trattner[4]

[1] Know-Center, Graz University of Technology, Graz, Austria
dkowald@know-center.at
[2] Knowledge Technologies Institute, Graz University of Technology, Graz, Austria
{simone.kopeinik,paul.seitlinger,dietrich.albert}@tugraz.at
[3] Institute of Informatics, Tallin University, Tallinn, Estonia
tley@tlu.ee
[4] Norwegian University of Science and Technology, Trondheim, Norway
chritrat@idi.ntnu.no

Abstract. In this paper, we introduce a tag recommendation algorithm that mimics the way humans draw on items in their long-term memory. Based on a theory of human memory, the approach estimates a tag's probability being applied by a particular user as a function of usage frequency and recency of the tag in the user's past. This probability is further refined by considering the influence of the current semantic context of the user's tagging situation. Using three real-world folksonomies gathered from bookmarks in BibSonomy, CiteULike and Flickr, we show how refining frequency-based estimates by considering usage recency and contextual influence outperforms conventional "most popular tags" approaches and another existing and very effective but less theory-driven, time-dependent recommendation mechanism.

By combining our approach with a simple resource-specific frequency analysis, our algorithm outperforms other well-established algorithms, such as FolkRank, Pairwise Interaction Tensor Factorization and Collaborative Filtering. We conclude that our approach provides an accurate and computationally efficient model of a user's temporal tagging behavior. We demonstrate how effective principles of recommender systems can be designed and implemented if human memory processes are taken into account.

Keywords: Personalized tag recommendations · Time-dependent recommender systems · Base-level learning equation · ACT-R · Human memory model · BibSonomy · CiteULike · Flickr

Parts of this work have been included as an extended version in the article "Modeling Activation Processes in Human Memory to Predict the Reuse of Tags" submitted to the *The Journal of Web Science*.

© Springer International Publishing Switzerland 2015
M. Atzmueller et al. (Eds.): MUSE/MSM 2013, LNAI 8940, pp. 55–74, 2015.
DOI: 10.1007/978-3-319-14723-9_4

1   Introduction

In this paper, we suggest a tag recommendation mechanism that mimics how people use to access their memory to name things they have encountered in the past. In everyday communication, people are very effective and quick in retrieving relevant knowledge from the enormous amount of information units stored in their individual long-term memory (LTM). One example is tagging resources on the Web, a rudimentary variant of communication [1,2]. Here, people name objects, such as images or music files, by means of social tags to create retrieval cues for personal and collective information organization [3]. The issue of how human memory ensures a fast and automatic information retrieval from its huge LTM has been extensively examined by memory psychology (e.g., [4]). Essentially, human memory is tuned to the statistical structure of an individual's environment and keeps available those memory traces that have been used frequently and recently in the past and are relevant in the current context [5].

Social tagging provides an illustrative example of the strong interplay between external, environmental and internal memory structures and processes (e.g., [6]). For instance, the development of generative models of social tagging demonstrated that the probability of a tag being applied can be modeled through the preferential attachment principle (e.g., [7]): the higher the frequency of a tag's past occurrence in the tagging environment is, the more likely it will be reused by an individual. Additionally, the same probability is also a function of the tag's recency, which is the time elapsed since the tag last occurred in the environment [8]. In summary, the probability of applying a particular word reflects the individual's probability of being exposed to the word in her environment [5].

The activation Eq. 1 of the cognitive architecture ACT-R (e.g., [4]) is an empirically well-established formula to estimate the activation A_i of a memory trace for an item i (e.g., the tag "recognition"), where the psychological construct of item activation is assumed to control for the item's retrieval from the LTM. It is given by:

$$A_i = B_i + \underbrace{\sum_j W_j \times S_{j,i}}_{Associative Component} \tag{1}$$

A wealth of empirical data (e.g., [4,9]) demonstrates that A_i is a function of the item's base-level activation B_i and associative activation caused by the associative component. B_i reflects the general usefulness of item i in the individual's past and is given by the base-level learning equation:

$$B_i = ln(\sum_{j=1}^{n} t_j^{-d}), \tag{2}$$

where n is the frequency of item occurrences in the past and t_j is the recency, which is the time since the jth occurrence. For example, if a user has applied the two tags "recognition" and "recommender" with equal frequency, i.e., for an equal number of bookmarks, but "recommender" has dominated the user's recent bookmarks, the equation predicts a higher activation and hence, larger probability of being reused, for "recommender" than for "recognition". The exponent d accounts for the power-law of forgetting and models the phenomenon that each memory's activation, caused by the jth occurrence, decreases in time according to a power function. The exponent d is typically set to 0.5 [4].

The second component of A_i, the associative activation, is assumed to adjust B_i according to the individual's current context that may consists of words included in a resource's abstract or popular tags assigned to the resource. For example, even if the base-level activation for the tag "recognition" is smaller than for "recommender", a particular set of contextual elements, such as the words "memory" and "recollection", will spread associative activation to "recognition" and will substantially increase its probability being reused. In Eq. 1, W_j represents the weights of the items j, which are elements of the current context (e.g., "memory" and "recollection"); $S_{j,i}$ represents the strengths of association from the contextual elements to an item (e.g., "recognition"). Section 3 gives a detailed and formal description of how these two components are calculated.

In the present work, we test the assumption that the two components of Eq. 1 (time and semantic context) can be used to improve frequency-based tag reuse predictions. Specifically, we raise the following three research questions:

- *RQ1*: Does the base-level learning equation provide a valid model of a user's tagging behavior in the past to predict future tag assignments?

- *RQ2*: Does the additional consideration of the associative component evoked by the current context further improve the accuracy of the base-level learning equation?

- *RQ3*: Can the whole activation equation, that considers base-level and associative activation, be applied and extended to create an effective and efficient tag recommendation mechanism compared to state-of-the-art baseline approaches?

The strategy we chose to address all research questions consists of three steps. In a first step, we implemented the "pure" BLL equation in form of a tag recommender and compared its performance with a MostPopular$_u$ (MP$_u$) approach, which suggests the most frequent tags in a user's tag assignments. As expected, the comparison with MP$_u$ showed evidence of the incremented value that results when additionally processing the recency of tag use. Moreover, we compared our BLL recommender with the currently leading time-based tag recommender approach introduced by [10] and showed the advantages of our theory-driven approach.

In a second step, we extended the "pure" BLL equation to the full activation equation proposed by Anderson et al. [4]. In this way, we also take into account relevant context information (i.e., tags applied to the target resource) and fine-tune base-level activation values. This has led to an improvement over the "pure" BLL equation, and showed particularly good results in settings where context information is important (e.g., if there is a high probability that previously assigned resource tags get adapted by the users).

In a third step, we combined the activation equation with popular tags that have been applied to the target resource by other users. When also considering other users' tags, it allows us to introduce new tags to the target user, namely tags that have not been used by the target user before (e.g., [11,12] or [13]). To this end we weighted the tags based on their frequency in the resource's tag assignments, hereinafter referred to as MostPopular$_r$ (MP$_r$). We then compared the performance of the combination of the activation equation and MP$_r$ with well-established approaches, such as Collaborative Filtering (CF), FolkRank (FR) and Pairwise Interaction Tensor Factorization (PITF), and showed that this approach outperforms the state-of-the-art algorithms in terms of recommender accuracy.

The remainder of this paper is organized as follows: we begin with discussing related work (Sect. 2) and describing our approach in Sect. 3. Sections 4 and 5 describe the experimental setup and the baseline algorithms we used for our evaluation. Section 6 addresses our three research questions and summarizes the settings and results of our extensive evaluation. Finally, in Sect. 7, we conclude the paper by discussing our findings in the light of the benefits of deriving tag recommender mechanisms from empirical, cognitive research.

2   Related Work

Recent years have shown that tagging is an important feature of the Social Web supporting the users with a simple mechanism to collaboratively organize and find content [14]. Although tagging has the ability to improve search (in particular tags provided by the individual) [15,16], it is also known that users are typically lazy in providing tags for instance for their bookmarked resources. It is therefore not surprising that recent research has attempted to address this challenge to support the individual in her tag application process in the form of personalized tag recommenders. To date, the two following approaches have been established: graph-based and content-based tag recommender systems [12]. In our work we focus on graph-based approaches.

The probably most notable work in this context is the work of Hotho et al. [17] who introduced an algorithm termed FolkRank (FR) that has become the most prominent benchmarking tag recommender approach over the past few years. Subsequently, the work of Jäschke et al. [18] and Hamouda & Wanas [19] showed how the classic Collaborative Filtering (CF) approach could be adopted for the problem of predicting tags to the user in a personalized manner. More recent work in this context are studies of Rendle et al. [20], Wetzker et al. [21], Krestel et al. [22] or Rawashdeh et al. [23] who introduced a factorization model, a Latent Dirichlet Allocation (LDA) model or a Link-Prediction model, based on the Katz measure, to recommend tags to users.

Within the context of this paper, another relevant study addressing the potential of social tagging systems to model the user in a resource context is presented in [24]. In this work similarities between users are defined through firstly, their consensus in tagging behavior and secondly, their resource usage. The resulting network with actors, resources, tags and tag assignments as nodes, is modeled as a directed graph. The graph serves as a basis to spread activation from one actor to another, by going through multiple types of nodes that represent context information linking the actors. Stanley et al. [25] studied a tag recommendation model inspired by the declarative memory retrieval mechanism of ACT-R (e.g., [4]) on forum data. Recommendations are based on two aspects, the user's tag history and co-occurrences between tag words and words extracted from the post's content. Although the model initially aims to implement the entire activation equation of ACT-R, since it was tailored to the properties of a computer science forum, where assumingly the frequency of tags does not change significantly over time, the approach does not consider the time component as it is done in this paper. Sigurbjörnsson et al. [26] suggests a slightly different approach to calculate relatedness values between tags in order to recommend tags in the image-sharing portal Flickr. The proposed method is based on the Jaccard coefficient to normalize the co-occurrence of two tags

Although the mentioned approaches perform reasonably well, they are computational expensive compared to simple "most popular tags" approaches. Furthermore, they ignore recent observations with regard to social tagging systems, such as the variation of the individual tagging behavior over time [27]. To that end, recent research has made the first promising steps towards more accurate graph-based models that also account for the variable of time [10,28]. The approaches have shown to outperform some of the current state-of-the-art tag recommender algorithms.

In line with the latter strand of research, in this paper we present a novel graph-based tag recommender mechanism that uses the activation equation that is part of the former mentioned ACT-R theory (e.g., [4]) to integrate item frequency and recency as well as current context information. We show that the approach is not only very simple and straightforward but also reveal that the algorithm outperforms current state-of-the-art graph-based (e.g., [17,18,21]) and the leading time-based [10] tag recommender approaches.

3 Approach

In Sect. 1 we formulated the assumption that the activation equation consisting of a base-level and associative component can be applied to explain a high variance in a tag's probability of being applied. To address our first research question (as to whether the activation equation's first component can be applied to improve a "most popular tags by user" approach) we have calculated the base-level activation $B(t, u)$ of a given tag t in a user u's set of tag assignments, $Y_{t,u}$. First, we determined a reference timestamp $timestamp_{ref}$ (in seconds) that is the timestamp of the most recent bookmark of user u. In our dataset samples, $timestamp_{ref}$ corresponded to the timestamp of the user's bookmark that was selected for the test set (see Sect. 4.2).

If $i = 1 \ldots n$ indexes all tag assignments in $Y_{t,u}$, the recency of a particular tag assignment is given by $timestamp_{ref} - timestamp_i$. Finally, $B(t, u)$ of tag t for a user u is given by the BLL equation:

$$B(t, u) = \ln(\sum_{i=1}^{n} (timestamp_{ref} - timestamp_i)^{-d}), \qquad (3)$$

where d is set to 0.5 based on [4]. In order to map the values onto a range of 0 to 1 we applied a normalization method as proposed in related work [29]:

$$\|B(t, u)\| = \frac{\exp(B(t, u))}{\sum\limits_{t' \in Y_u} \exp(B(t', u))}, \qquad (4)$$

where Y_u is the set of unique tags used by user u in the past.

To investigate our second research question (as to whether the BLL equation can be further improved by also considering the associative component evoked by the current context) we have implemented Eq. 1 in form of:

$$A(t, u) = \|B(t, u)\| + \underbrace{\sum_j W_j \times S_{j,i}}_{AssociativeComponent} \qquad (5)$$

To calculate the variables of the associative component, i.e., to model a user's semantic context, we simply looked at the tags assigned by other users to the given resource r. A user's semantic context certainly consists of much more aspects, such as content words in the title or in the page text. However, since not all of our datasets contain title information or page text and other studies have convincingly demonstrated the impact of a resource's prominent tags on a user's tagging behavior (e.g., [11,12]), we decided to approximate the context by means of other users' tags.

When applying the formula to a recommender system, related literature [26,30] suggests to use a measure of normalized tag co-occurrence to represent the strength of an association. Accordingly, we define the co-occurrence between two tags as the number of bookmarks in which both tags are included. To add meaning to the co-occurrence value, the overall frequency of the two tags is also taken into consideration. This is done by normalizing the co-occurrence value according to the Jaccard coefficient (6) following the approach described in [26]:

$$S_{j,i} = \frac{|t_j \cap t_i|}{|t_j \cup t_i|} \qquad (6)$$

In our implementation, $S_{j,i}$ is calculated as an association value between a tag previously given by the target user (t_i) and a tag that has been assigned to a resource of interest (t_j). Based on a tag co-occurence matrix that depicts the tag relations of an entire data set, information about how many times two tags co-occur ($t_j \cap t_i$) is retrieved and set into relation with the number of bookmarks in which at least one of the two tags appear $(t_j \cap t_i)$. We set the attentional weight W_j to the number of times t_j occurred in the tag assignments of the target resource.

Finally, to examine our third research question (as to whether the activation equation can be implemented in form of an effective recommender mechanism) we extended Eq. 5 by also considering the most popular tags in the tag assignments of the resource Y_r (MP_r, i.e., $arg\, max\, t \in T(|Y_r|)$) [17]. This simple extension was necessary to allow for the prediction of new and plausible tags that a user has not assigned in her or his previous tagging history (e.g., [11,12] or [13]). Finally, the list of recommended tags for a given user u and resource r is calculated by the following equation:

$$\widetilde{T}(u,r) = arg\, max\, t \in T(\underbrace{\beta\, \|A(t,u)\|}_{BLL_{AC}} + (1-\beta)\|\|Y_{t,r}\|\|), \qquad (7)$$
$$\underbrace{\qquad\qquad\qquad\qquad\qquad\qquad\qquad\qquad}_{BLL_{AC}+MP_r}$$

where β is used to inversely weight the two components, i.e. the activation values $A(t,u)$ and the most popular tags of the target resource given by MP_r. The results presented in Sect. 6 were calculated using $\beta = 0.5$. However, we focused on the performance of $BLL_{AC}+MP_r$ in the experiments, i.e. on an approach estimating a tag's probability of being applied by means of user and corresponding resource information. The source-code of our approaches [31] is open-source and can be found online[1].

4   Experimental Setup

In this section we describe in detail the datasets, the evaluation method and the metrics used in our experiments.

4.1   Datasets

For the purposes of our study and for reasons of reproducibility, we focused our investigations on three well-known and freely-available folksonomy datasets. To test our approach on both types of advocates, known as broad and narrow folksonomies [32] (in a broad folksonomy many users are allowed to annotate a particular resource while in a narrow folksonomy only the user who has uploaded the resource is permitted to apply tags), freely available datasets from the social bookmark and publication sharing system BibSonomy[2], the reference management system CiteULike[3] (broad folksonomies) and the image and video sharing platform Flickr[4] (narrow folksonomy) were utilized.

---

[1] https://github.com/learning-layers/TagRec/.
[2] http://www.kde.cs.uni-kassel.de/bibsonomy/dumps.
[3] http://www.citeulike.org/faq/data.adp.
[4] http://www.tagora-project.eu/.

Table 1. Properties of the datasets, where $|B|$ is the number of bookmarks, $|U|$ the number of users, $|R|$ the number of resources, $|T|$ the number of tags and $|TAS|$ the number of tag assignments.

Dataset	Core	$|B|$	$|U|$	$|R|$	$|T|$	$|TAS|$
BibSonomy	-	400,983	5,488	346,444	103,503	1,479,970
	3	41,764	788	8,711	5,757	161,509
CiteULike	-	3,879,371	83,225	2,955,132	800,052	16,703,839
	3	735,292	17,983	149,220	67,072	2,242,849
Flickr	-	864,679	9,590	864,679	127,599	3,552,540
	3	860,135	8,332	860,135	58,831	3,465,346

Since automatically generated tags affect the performance of the tag recommender systems, we excluded all of those tags from the datasets (e.g., for BibSonomy and CiteULike we excluded the *no-tag*, *bibtex-import*-tag, etc.). Furthermore, we decapitalized all tags as suggested by related work in the field (e.g., [20]). In the case of Flickr we randomly selected 3 % of the user profiles for reasons of computational effort (see also [33]). The overall dataset statistics can be found in Table 1. As can be seen in column "Core", we applied both: a p-core pruning approach [34] (represented by "3") to capture the issues of data sparseness, as well as no p-core pruning (represented by "-") to capture the issue of cold-start users or items [35], respectively.

4.2   Evaluation Methodology

To evaluate our tag recommender approach we used a leave-one-out hold-out method as proposed by popular and related work in this area (e.g., [18]). Hence, we created two sets, one set for training and the other set for testing. To split up each dataset in two, we eliminated for each user her latest bookmark (in time) from the original dataset and added it to the test set. Each bookmark in the test set consists of a collection of one or more tags to which we further refer as relevant tags. The reduced original dataset was used for training, and the newly created one for testing. This procedure simulates a real-world environment well and is a recommended offline-evaluation procedure for time-based recommender systems [36]. To finally quantify the performance of our approaches, a set of well-known information retrieval performance standard metrics were utilized [12,18]:

Recall (R@k) is calculated as the number of correctly recommended tags divided by the number of relevant tags, where t_u^k denotes the top k recommended tags and T_u the list of relevant tags of a bookmark of user u:

$$R@k = \frac{1}{|U|} \sum_{u \in U} \frac{|t_u^k \cap T_u|}{|T_u|} \tag{8}$$

Precision (P@k) is calculated as the number of correctly recommended tags divided by the number of recommended tags. As it can be seen in the following formula, in contrast to R@k, P@k usually decreases with a higher number of recommended tags k:

$$P@k = \frac{1}{|U|} \sum_{u \in U} \frac{|t_u^k \cap T_u|}{|t_u^k|} \tag{9}$$

F1-score (F1@k) combines recall and precision into one score. It is calculated as the product of P@k and R@k divided by the sum of R@k and P@k multiplied by 2 [12]:

$$F1@k = \frac{1}{|U|} \sum_{u \in U} (2 \times \frac{R@k \times P@k}{R@k + P@k}) \tag{10}$$

Mean Reciprocal Rank (MRR) is a ranking-dependent metric and is calculated as the sum of the reciprocal ranks of all relevant tags in the list of the recommended tags. This means that a higher MRR is achieved if the relevant tags occur at the beginning of the recommended tag list [23]:

$$MRR = \frac{1}{|U|} \sum_{u=1}^{|U|} (\frac{1}{|T_u|} \sum_{t \in T_u} \frac{1}{rank(t)}) \tag{11}$$

Mean average precision (MAP) is an extension of the precision metric that also looks on the ranking of the recommended tags. It is described in the formula below where B_k is 1 if the recommended tag at position k is relevant [23].

$$MAP = \frac{1}{|U|} \sum_{u=1}^{|U|} (\frac{1}{|T_u|} \sum_{k=1}^{|t_u^k|} B_k \times P@k) \tag{12}$$

In particular, we report R@k, P@k, MRR and MAP for $k = 10$ and F1-Score (F_1@k) for $k = 5$ recommended tags[5].

5   Baseline Algorithms

We compared our approaches to several baseline tag recommender algorithms. The algorithms range from simple frequency-based approaches to more complex mechanisms based on factorization or temporal models and have been selected based on their popularity in the community, performance and novelty [37,38]. The used baselines are the following:

---

[5] F_1@5 was also used as the main performance metric in the ECML PKDD Discovery Challenge 2009: http://www.kde.cs.uni-kassel.de/ws/dc09/.

MostPopular (MP). This approach is an unpersonalized tag-recommender algorithm that does not take into account information about the target user or resource. MP recommends for any user and any resource the same set of tags that is weighted by the frequency in all tag assignments [39].

MostPopular$_u$ (MP$_u$). The *most popular tags by user* approach processes tagging information about the user but ignores the resource component, which means that a target user receives the same personalized tag suggestions, no matter which resource is going to be tagged. MP$_u$ suggests the most frequent tags in the tag assignments of the user [39].

MostPopular$_r$ (MP$_r$). The *most popular tags by resource* algorithm uses the previously assigned resource tags for the prediction process but ignores information about the target user. MP$_r$ weights the tags based on their frequency in the tag assignments of the resource [39].

MostPopular$_{u,r}$ (MP$_{u,r}$). This algorithm is a mixture of the most popular tags by user (MP$_u$) and most popular tags by resource (MP$_r$) approaches. MP$_{u,r}$ joins both components using a simple linear combination [18].

Collaborative Filtering (CF). Marinho et al. [40] described how the classic Collaborative Filtering (CF) approach [41] can be adapted for tag recommendations. Since folksonomies have ternary relations (users, resources and tags), the classic CF approach can not be applied directly. Thus, the neighborhood of an user is formed based on the tag assignments in the user profile. Furthermore, in CF-based tag recommendations only the subset of users that have tagged the target resource is taken into account when calculating the user neighborhood (*Note*: if there are no users that have tagged the target resource in the past, all users are treated as candidates for the neighborhood). The set of recommended tags can then be determined based on this neighborhood [18,40]. We used a neighborhood size of 20 as suggested in related work [33].

Adapted PageRank (APR). Hotho et al. [17] adapted the well-known PageRank algorithm in order to rank the nodes within the graph structure of a folksonomy. This is based on the idea that a resource is important if it is tagged with important tags by important users. Thus, the folksonomy is converted into an undirected graph, where the co-occurences of users and resources, users and tags and resources and tags are treated as weighted edges.

FolkRank (FR). The FolkRank algorithm is an extension of the Adapted PageRank approach that was also proposed by Hotho et al. [17]. This extension gives a higher importance to the preference vector via a differential approach [18]. Our APR and FR implementations are based on the code and settings of the open-source Java tag recommender framework provided by the University of Kassel[6].

Factorization Machines (FM). Rendle [42] introduced Factorization Machines which combine the advantages of Support Vector Machines (SVM) with factorization models to build a general prediction model that is also capable of tag recommendations. In contrast to SVMs, FMs are able to estimate interactions between entities even in settings with huge sparsity (e.g., recommender systems).

Pairwise Interaction Tensor Factorization (PITF). This approach proposed by Rendle and Schmidt-Thieme [20] is an extension of Factorization Machines and is based on the Tucker Decomposition (TD) model. In contrast to TD, PITF explicitly models the pairwise interactions between users, resources and tags to provided personalized tag recommendations. The FM and PITF results presented in this paper were calculated using the open-source C++ tag recommender framework provided by the University of Konstanz[7] with 256 factors, as suggested by [20].

Temporal Tag Usage Patterns (GIRP). This time-dependent tag recommender algorithm was presented by Zhang et al. [10] and is based on the frequency and the temporal usage of a user's tag assignments. In contrast to BLL it models the temporal tag usage with an exponential distribution rather than a power-law distribution.

GIRP with Tag Relevance to Resource (GIRPTM). This is an extension of the GIRP algorithm that also takes the resource component into account [10]. This is done in the same manner as in $BLL_{AC}+MP_r$, thus adding the probability values of MP_r using a simple linear combination.

6   Results

The presentation of the evaluation results is organized in line with our three research questions, as introduced in Sect. 1. With respect to the recommender quality, we will turn our attention first to the BLL equation and its validity to model tagging behavior, second, to the impact of context information when added to the BLL equation (BLL_{AC}) and third, to a comparison of our context enriched BLL implementation $(BLL_{AC} + MP_r)$ with state-of-the-art baseline approaches.

   The BLL equation models the user's tagging behavior with respect to frequency and recency. While the frequency of tag use is a fairly common parameter for tag recommendations, the factor of time, that models the effects of a user's long term memory (as described through recency), is expected to bring

---

[6] http://www.kde.cs.uni-kassel.de/code.
[7] http://www.informatik.uni-konstanz.de/rendle/software/tag-recommender/.

Table 2. F_1@5, MRR and MAP values for BibSonomy, CiteULike and Flickr (no core and core 3) showing that the BLL equation provides a valid model of a user's tagging behavior to predict tags (first research question). Moreover, the results imply that using the activation equation (BLL$_{AC}$) to also take into account semantic cues (i.e., associations with resource tags) can further improve this model (second research questions).

Dataset	Core	Measure	MP$_u$	GIRP	BLL	BLL$_{AC}$
BibSonomy	-	F_1@5	.152	.157	.162	**.169**
		MRR	.114	.119	.125	**.133**
		MAP	.148	.155	.162	**.172**
	3	F_1@5	.215	.221	.228	**.292**
		MRR	.202	.210	.230	**.286**
		MAP	.238	.247	.272	**.345**
CiteULike	-	F_1@5	.185	.194	.201	**.211**
		MRR	.165	.182	.193	**.205**
		MAP	.194	.213	.227	**.242**
	3	F_1@5	.272	.291	.300	**.336**
		MRR	.268	.294	.319	**.365**
		MAP	.305	.337	.366	**.424**
Flickr	-	F_1@5	.435	.509	.523	**.523**
		MRR	.360	.445	.466	**.466**
		MAP	.468	.590	.619	**.619**
	3	F_1@5	.488	.577	.592	**.592**
		MRR	.407	.511	.533	**.533**
		MAP	.527	.676	.707	**.707**

additional value to tag recommendation approaches. That is why we investigate our first research question by determining the effect of the recency component on tag assignments.

When comparing BLL with MP$_u$ and GIRP, the results reported in Table 2 and Fig. 1 clearly show that the time-dependent algorithms BLL and GIRP both outperform the frequency-based MP$_u$ approach. Looking further at the two time-dependent algorithms, BLL reaches higher levels of accuracy than the less theory-driven GIRP algorithm in both settings (with and without p-core pruning). Even more apparent is the impact of the recency component in the narrow folksonomy (Flickr). Unlike the broad folksonomies (BibSonomy, CiteULike), the Flickr dataset has no tags of other users available for the target resource. Therefore, a user needs to assign tags without having the inspiration of previously given tags. We assume that the user, to this end, needs to draw on her long term memory that the BLL equation aims to mimic. In summary, these results prove that the BLL equation provides a valid model of a user's tagging behavior to predict tags (first research question).

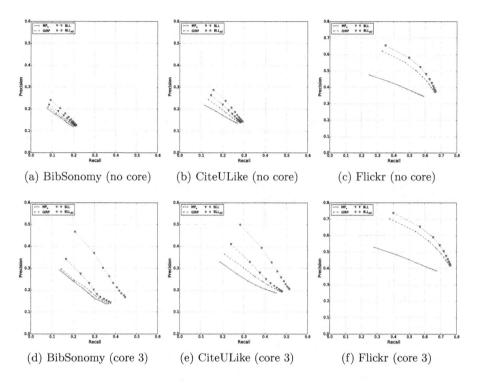

Fig. 1. Recall/Precision plots for BibSonomy, CiteULike and Flickr (no core and core 3) showing the performance of BLL and BLL_{AC} along with MP_u and GIRP for 1 - 10 recommended tags (k).

By expanding BLL to BLL_{AC}, we implement the activation equation as explained in Sect. 3. The activation equation enriches the base-level activation (i.e., frequency and recency of tag use) by adding contextual activation through tags previously assigned to the target resource. Looking at the results of this experiment, as illustrated in Table 2 and Fig. 1, a number of interesting aspects appear. For one thing, it demonstrates that BLL_{AC} reveals only a small improvement over BLL, when applied on the unfiltered datasets (no p-core) of the broad folksonomies (BibSonomy and CiteULike). However, this changes when looking at the results for the p-core pruned datasets (core 3). Caused by the higher number of tags assigned to each resource, the contextual activation gains impact. This leads to considerably increased values for all of the used metrics ($F1@5$, MRR, MAP). One might wonder why the results of BLL and BLL_{AC} are the same in the case of the narrow folksonomy (Flickr). This is, in fact, an expected outcome. As resources in the Flickr dataset are tagged by only one user (i.e., the one that has uploaded it), the model of the resource component does not generate additional value. According to these results we can answer the second research question positively, since the fine-tuning or re-ranking of the user tags based on context cues increases the recommender accuracy in the broad folksonomies BibSonomy and CiteULike.

To address our third and last research question we combine our BLL_{AC} approach with MP_r, which leads to $BLL_{AC}+MP_r$. Hereby, BLL_{AC} models the context-aware user component while MP_r further models the resource component in order to also take into account new tags that have not been used by the target user in the past. The results presented in Table 3 show that this approach outperforms a set of state-of-the-art baseline algorithms as well as $BLL+MP_r$ (without contextual activation of the user tags). Further looking into the results, it becomes apparent that the three time-dependent algorithms (GIRPTM, $BLL+MP_r$ and $BLL_{AC}+MP_r$) produce higher estimates ($F_1@5$, MRR and MAP) across all datasets as well as in both settings (with and without p-core pruning). Moreover, an important observation is that our $BLL_{AC}+MP_r$ approach also outperforms GIRPTM, the currently leading, graph-based time-depended tag recommendation algorithm. Particularly good results are shown here for the ranking-dependent metrics such as MRR and MAP. This observation clearly shows the advantages of our approach, that is build upon long-standing models of human memory theory, over the less-theory driven GIRPTM algorithm.

Another aspect worth discussing is the contrast of the results illustrated in Table 2, where BLL_{AC} reaches substantially higher levels of accuracy than BLL, to the results outlined in Table 3, where $BLL_{AC}+MP_r$ only indicate marginal improvements over $BLL+MP_r$. In our opinion, this effect appears because the resource tag information depicted in MP_r is congruent with data used for the contextual activation in BLL_{AC}. This finding suggests that the use of other resource metadata, such as title or body-text, may be valuable when specifying the context in BLL_{AC} (see also Sect. 7). Similar observations can be made when looking at the Recall/Precision curves in Fig. 2.

In summary, our results clearly imply that the activation equation by Anderson et al. [4] can be used to implement a highly effective recommender approach. Overall, the simulations demonstrate that it exceeds the performance of well-established and effective recommenders, such as $MP_{u,r}$, CF, APR, FM and even the currently leading time-dependent approach GIRPTM [10] (third research question). Finally, it is indispensable to highlight that $BLL_{AC}+MP_r$, despite its simplicity, appears to be even more successful than the sophisticated FR and PITF algorithms.

7   Discussion and Conclusion

In this study we followed a three-step strategy. We started by comparing the performance of BLL with MP_u to determine the effect of considering the recency of each tag use. Results of an additional comparison may differentiate our cognitive-psychological model from the less theory-driven GIRP approach introduced by Zhang et al. [10]. Our findings, tackling the first research question, clearly demonstrate that regardless of the evaluation metric and across all datasets, BLL reaches higher levels of accuracy than MP_u and even outperforms GIRP. Thus, processing the recency of tag use is effective to account for additional variance

Table 3. $F_1@5$, MRR and MAP values for BibSonomy, CiteULike and Flickr (no core and core 3) showing that our $BLL_{AC}+MP_r$ approach outperforms state-of-the-art baseline algorithms (third research question).

Dataset	Core	Measure	MP	MP_r	$MP_{u,r}$	CF	APR	FR	FM	PITF	GIRPTM	BLL+MP_r	$BLL_{AC}+MP_r$
BibSonomy	-	$F_1@5$	.013	.074	.192	.166	.175	.171	.122	.139	.197	.201	**.202**
		MRR	.008	.054	.148	.133	.149	.148	.097	.120	.152	.158	**.159**
		MAP	.009	.070	.194	.173	.193	.194	.120	.150	.200	.207	**.209**
	3	$F_1@5$	.047	.313	.335	.325	.260	.337	.345	.356	.350	.353	**.358**
		MRR	.035	.283	.327	.289	.279	.333	.329	.341	.334	.349	**.350**
		MAP	.038	.345	.403	.356	.329	.414	.408	.421	.416	.435	**.439**
CiteULike	-	$F_1@5$	.002	.131	.253	.218	.195	.194	.111	.122	.263	.270	**.271**
		MRR	.001	.104	.229	.201	.233	.233	.110	.141	.246	.258	**.259**
		MAP	.001	.134	.280	.247	.284	.284	.125	.158	.301	.315	**.317**
	3	$F_1@5$	.013	.270	.316	.332	.313	.318	.254	.258	.336	.346	**.351**
		MRR	.012	.243	.353	.295	.361	.366	.282	.290	.380	.409	**.415**
		MAP	.012	.294	.420	.363	.429	.436	.326	.334	.455	.489	**.497**
Flickr	-	$F_1@5$	.023	-	.435	.417	.328	.334	.297	.316	.509	.523	**.523**
		MRR	.023	-	.360	.436	.352	.355	.300	.333	.445	.466	**.466**
		MAP	.023	-	.468	.581	.453	.459	.384	.426	.590	.619	**.619**
	3	$F_1@5$	.026	-	.488	.493	.368	.378	.361	.369	.577	.592	**.592**
		MRR	.026	-	.407	.498	.398	.404	.375	.390	.511	.533	**.533**
		MAP	.026	-	.527	.663	.513	.523	.481	.502	.676	.707	**.707**

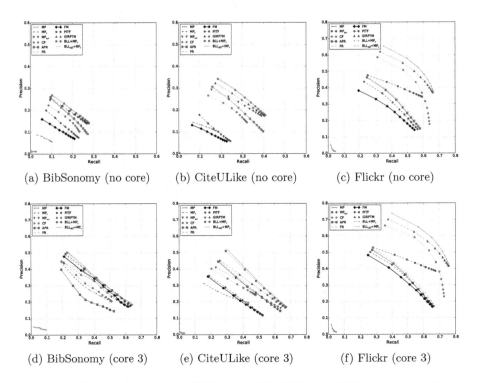

Fig. 2. Recall/Precision plots for BibSonomy, CiteULike and Flickr (no core and core 3) showing the performance of $BLL+MP_r$ and $BLL_{AC}+MP_r$ along with state-of-the-art baseline mechanisms for 1 - 10 recommended tags (k)

of users' tagging behavior and therefore, a reasonable extension of simple "most popular tags" approaches. Furthermore, the advantage over GIRP indicates that drawing on memory psychology guides the application of a reliable and valid model built upon long-standing, empirical research. The equations that Zhang et al. [10] used to implement their approach were developed from scratch rather than derived from existing research described above. As a consequence, [10] models the recency of tag use by means of an exponential function, which is clearly at odds with the power law of forgetting described in related work (e.g., [5]).

In a second step, we have extended BLL to BLL_{AC} using current context information based on the activation equation of Anderson et al. [4]. Where BLL gives the prior probability of tag reuse that is learned over time, the associative component tunes this prior probability to the current context by exploiting the current semantic cues from the environment (i.e., the previously assigned tags of the target resource). This is in line with how ACT-R models the retrieval from long-term memory. Our results show that this step clearly improves the "pure" BLL equation, especially in case of the p-core pruned datasets, where more context information (i.e., tag assignments of the target resource) are available to calculate the associative component.

In a third step, we combined BLL$_{AC}$ with the most popular tags that have been applied by other users to the target resource in the past (i.e., MP$_r$) in order to be able to also recommend new tags, i.e. tags that have not been used by the target user before. Despite their simplicity, our results show that this combination (BLL$_{AC}$+MP$_r$) has potential to outperform well-established mechanisms, such as CF, FR and PITF. We assume this is the case because, in following some fundamental principles of human memory, our approaches are better adapted to the statistical structure of the environment. Moreover, the results of this experiment also show that there is only a small difference between BLL$_{AC}$+MP$_r$ and BLL+MP$_r$ (without contextual activation of the user tags), which suggests the use of additional context information, such as content-based features (e.g., the resource's title or body-text). This would also be in line with the study presented in [43], where the authors show that the resource title has a big impact on tags in collaborative tagging systems and so could be a better alternative to represent context cues than the popular tags of the resource used in the current work.

Finally, a glance on the results shows an interdependency between the examined dataset and the performance of our approaches. While the distance to other strongly performing mechanisms does not appear to be large in case of broad folksonomies (BibSonomy and CiteULike), this distance gets substantially larger in a narrow folksonomoy (Flickr), where no tags of other users are available for the target user's resources. From this interdependency we conclude that applying a model of human memory is primarily effective if tag assignments are mainly driven by individual habits unaffected by the behavior of other users, such as it is done in Flickr.

In future work, we will continue examining memory processes that are involved in categorizing and tagging Web resources. For instance, in a recent study [44], we introduced a mechanism by which memory processes involved in tagging can be modeled on two levels of knowledge representation: on a semantic level (representing categories or LDA topics) and on a verbal level (representing tags). Next, we will aim at combining this integrative mechanism with the activation equation to examine a potential correlation between the impact of recency (time-based forgetting) and the level of knowledge representation. We believe that conclusions drawn from cognitive science will help to develop an effective and psychologically plausible tag recommendation mechanism.

Finally, we plan to compare our approach also to other baselines, e.g., the hybrid tag recommender algorithm mentioned in [45], and to use more datasets, e.g., LastFm, to further show that our approach is capable of capturing various kinds of tagging behavior.

Acknowledgments. This work is supported by the Know-Center, the EU funded projects Learning Layers (Grant Agreement 318209) and weSPOT (Grant Agreement 318499) and the Austrian Science Fund (FWF): P 25593-G22. Moreover, parts of this work were carried out during the tenure of an ERCIM "Alain Bensoussan" fellowship programme. The Know-Center is funded within the Austrian COMET Program - Competence Centers for Excellent Technologies - under the auspices of the Austrian Ministry of Transport, Innovation and Technology, the Austrian Ministry of Economics and Labor and by the State of Styria. COMET is managed by the Austrian Research Promotion Agency (FFG).

References

1. Halpin, H., Robu, V., Shepherd, H.: The complex dynamics of collaborative tagging. In: Proceedings of the 16th International Conference on World Wide Web. WWW '07, pp. 211–220. ACM, New York (2007)
2. Steels, L.: Semiotic dynamics for embodied agents. IEEE Intell. Syst. **21**, 32–38 (2006)
3. Marlow, C., Naaman, M., Boyd, D., Davis, M.: Ht06, tagging paper, taxonomy, flickr, academic article, to read. In: Proceedings of the Seventeenth Conference on Hypertext and Hypermedia. HYPERTEXT '06, pp. 31–40. ACM, New York (2006)
4. Anderson, J.R., Byrne, M.D., Douglass, S., Lebiere, C., Qin, Y.: An integrated theory of the mind. Psychol. Rev. **111**, 1036–1050 (2004)
5. Anderson, J.R., Schooler, L.J.: Reflections of the environment in memory. Psychol. Sci. **2**, 396–408 (1991)
6. Held, C., Kimmerle, J., Cress, U.: Learning by foraging: the impact of individual knowledge and social tags on web navigation processes. Comput. Hum. Behav. **28**, 34–40 (2012)
7. Dellschaft, K., Staab, S.: An epistemic dynamic model for tagging systems. In: Proceedings of the Nineteenth ACM Conference on Hypertext and Hypermedia. HT '08, pp. 71–80. ACM, New York (2008)
8. Cattuto, C., Loreto, V., Pietronero, L.: Semiotic dynamics and collaborative tagging. Proc. Natl. Acad. Sci. **104**, 1461–1464 (2007)
9. Pirolli, P.L., Anderson, J.R.: The role of practice in fact retrieval. J. Exp. Psychol. Learn. Mem. Cogn. **11**, 136 (1985)
10. Zhang, L., Tang, J., Zhang, M.: Integrating temporal usage pattern into personalized tag prediction. In: Sheng, Q.Z., Wang, G., Jensen, C.S., Xu, G. (eds.) APWeb 2012. LNCS, vol. 7235, pp. 354–365. Springer, Heidelberg (2012)
11. Lorince, J., Todd, P.M.: Can simple social copying heuristics explain tag popularity in a collaborative tagging system? In: Proceedings of the 5th Annual ACM Web Science Conference. WebSci '13, pp. 215–224. ACM, New York (2013)
12. Lipczak, M.: Hybrid tag recommendation in collaborative tagging systems. Ph.D. thesis, Dalhousie University (2012)
13. Kowald, D., Seitlinger, P., Trattner, C., Ley, T.: Long time no see: The probability of reusing tags as a function of frequency and recency. In: Proceedings of WWW '14, ACM, New York, (2014)
14. Körner, C., Benz, D., Hotho, A., Strohmaier, M., Stumme, G.: Stop thinking, start tagging: tag semantics emerge from collaborative verbosity. In: Proceedings of the 19th International Conference on World Wide Web. WWW '10, pp. 521–530. ACM, New York (2010)
15. Helic, D., Trattner, C., Strohmaier, M., Andrews, K.: Are tag clouds useful for navigation? a network-theoretic analysis. Int. J. Soc. Comput. Cyber-Phys. Syst. **1**, 33–55 (2011)
16. Trattner, C., Lin, Y.l., Parra, D., Yue, Z., Real, W., Brusilovsky, P.: Evaluating tag-based information access in image collections. In: Proceedings of the 23rd ACM Conference on Hypertext and Social Media, pp. 113–122. ACM (2012)

17. Hotho, A., Jäschke, R., Schmitz, C., Stumme, G.: Information retrieval in folksonomies: search and ranking. In: Sure, Y., Domingue, J. (eds.) ESWC 2006. LNCS, vol. 4011, pp. 411–426. Springer, Heidelberg (2006)
18. Jäschke, R., Marinho, L., Hotho, A., Schmidt-Thieme, L., Stumme, G.: Tag recommendations in folksonomies. In: Kok, J.N., Koronacki, J., Lopez de Mantaras, R., Matwin, S., Mladenič, D., Skowron, A. (eds.) PKDD 2007. LNCS (LNAI), vol. 4702, pp. 506–514. Springer, Heidelberg (2007)
19. Hamouda, S., Wanas, N.: Put-tag: personalized user-centric tag recommendation for social bookmarking systems. Soc. Netw. Anal. Min. **1**, 377–385 (2011)
20. Rendle, S., Schmidt-Thieme, L.: Pairwise interaction tensor factorization for personalized tag recommendation. In: Proceedings of the Third ACM International Conference on Web Search and Data Mining. WSDM '10, pp. 81–90. ACM, New York (2010)
21. Wetzker, R., Zimmermann, C., Bauckhage, C., Albayrak, S.: I tag, you tag: translating tags for advanced user models. In: Proceedings of the Third ACM International Conference on Web Search and Data Mining, pp. 71–80. ACM (2010)
22. Krestel, R., Fankhauser, P.: Language models and topic models for personalizing tag recommendation. In: 2010 IEEE/WIC/ACM International Conference on Web Intelligence and Intelligent Agent Technology (WI-IAT), vol. 1, pp. 82–89. IEEE (2010)
23. Rawashdeh, M., Kim, H.N., Alja'am, J.M., El Saddik, A.: Folksonomy link prediction based on a tripartite graph for tag recommendation. J. Intell. Inf. Syst. **40**, 1–19 (2012)
24. Troussov, A., Parra, D., Brusilovsky, P.: Spreading activation approach to tag-aware recommenders: modeling similarity on multidimensional networks. In: Proceedings of Workshop on Recommender Systems and the Social Web at the 2009 ACM Conference on Recommender Systems, RecSys, vol. 9 (2009)
25. Stanley, C., Byrne, M.D.: Predicting tags for stackoverflow posts. In: Proceedings of ICCM (2013)
26. Sigurbjörnsson, B., Van Zwol, R.: Flickr tag recommendation based on collective knowledge. In: Proceedings of the 17th International Conference on World Wide Web, pp. 327–336. ACM (2008)
27. Yin, D., Hong, L., Xue, Z., Davison, B.D.: Temporal dynamics of user interests in tagging systems. In: Twenty-Fifth AAAI Conference on Artificial Intelligence (2011)
28. Yin, D., Hong, L., Davison, B.D.: Exploiting session-like behaviors in tag prediction. In: Proceedings of the 20th International Conference Companion on World Wide Web, pp. 167–168. ACM (2011)
29. McAuley, J., Leskovec, J.: Hidden factors and hidden topics: Understanding rating dimensions with review text. In: Proceedings of the ACM Conference Series on Recommender Systems, ACM, New York (2013)
30. Van Maanen, L., Marewski, J.N.: Recommender systems for literature selection: A competition between decision making and memory models. In: Proceedings of the 31st Annual Conference of the Cognitive Science Society, pp. 2914–2919 (2009)
31. Kowald, D., Lacic, E., Trattner, C.: Tagrec: Towards a standardized tag recommender benchmarking framework. In: Proceedings of HT'14, ACM, New York (2014)
32. Helic, D., Körner, C., Granitzer, M., Strohmaier, M., Trattner, C.: Navigational efficiency of broad vs. narrow folksonomies. In: Proceedings of the 23rd ACM Conference on Hypertext and Social Media. HT '12, pp. 63–72. ACM, New York (2012)

33. Gemmell, J., Schimoler, T., Ramezani, M., Christiansen, L., Mobasher, B.: Improving folkrank with item-based collaborative filtering. In: Proceedings of theWorkshop on Recommender Systems and the Social Web (RSWEB '09), pp. 17–24. New York, NY, USA (2009)

34. Batagelj, V., Zaveršnik, M.: Generalized cores. arXiv preprint cs/0202039 (2002)

35. Doerfel, S., Jäschke, R.: An analysis of tag-recommender evaluation procedures. In: Proceedings of the 7th ACM Conference on Recommender Systems. RecSys '13, pp. 343–346. ACM, New York (2013)

36. Campos, P.G., Díez, F., Cantador, I.: Time-aware recommender systems: a comprehensive survey and analysis of existing evaluation protocols. User Model. User-Adapt. Interact. **24**, 1–53 (2013)

37. Marinho, L., Nanopoulos, A., Schmidt-Thieme, L., Jäschke, R., Hotho, A., Stumme, G., Symeonidis, P.: Social tagging recommender systems. In: Ricci, F., Rokach, L., Shapira, B., Kantor, P.B. (eds.) Recommender Systems Handbook, pp. 615–644. Springer, New York (2011)

38. Marinho, L.B., Hotho, A., Jäschke, R., Nanopoulos, A., Rendle, S., Schmidt-Thieme, L., Stumme, G., Symeonidis, P.: Recommender Systems for Social Tagging Systems. SpringerBriefs in Electrical and Computer Engineering. Springer, New York (2012)

39. Jäschke, R., Marinho, L., Hotho, A., Schmidt-Thieme, L., Stumme, G.: Tag recommendations in social bookmarking systems. AI Commun. **21**, 231–247 (2008)

40. Marinho, L.B., Schmidt-Thieme, L.: Collaborative tag recommendations. In: Preisach, C., Burkhardt, H., Schmidt-Thieme, L., Decker, R. (eds.) Data Analysis, Machine Learning and Applications, pp. 533–540. Springer, Heidelberg (2008)

41. Schafer, J.B., Frankowski, D., Herlocker, J., Sen, S.: Collaborative filtering recommender systems. In: Brusilovsky, P., Kobsa, A., Nejdl, W. (eds.) The Adaptive Web 2007. LNCS, vol. 4321, pp. 291–324. Springer, Heidelberg (2007)

42. Rendle, S.: Factorization machines. In: 2010 IEEE 10th International Conference on Data Mining (ICDM), pp. 995–1000. IEEE (2010)

43. Lipczak, M., Milios, E.: The impact of resource title on tags in collaborative tagging systems. In: Proceedings of the 21st ACM Conference on Hypertext and Hypermedia. HT '10, pp. 179–188. ACM, New York (2010)

44. Seitlinger, P., Kowald, D., Trattner, C., Ley, T.: Recommending tags with a model of human categorization. In: The ACM International Conference on Information and Knowledge Managament, ACM, New York (2013)

45. Gemmell, J., Schimoler, T., Mobasher, B., Burke, R.: Hybrid tag recommendation for social annotation systems. In: Proceedings of the 19th ACM International Conference on Information and Knowledge Management, pp. 829–838. ACM (2010)

Forgetting the Words but Remembering the Meaning: Modeling Forgetting in a Verbal and Semantic Tag Recommender

Dominik Kowald[1,2]($\boxtimes$), Paul Seitlinger[2], Simone Kopeinik[2], Tobias Ley[3], and Christoph Trattner[4]

[1] Know-Center, Graz University of Technology, Graz, Austria
dkowald@know-center.at

[2] Knowledge Technologies Institute, Graz University of Technology, Graz, Austria
{paul.seitlinger,simone.kopeinik}@tugraz.at

[3] Institute of Informatics, Tallin University, Tallinn, Estonia
tley@tlu.ee

[4] Norwegian University of Science and Technology, Trondheim, Norway
chritrat@idi.ntnu.no

Abstract. We assume that recommender systems are more successful, when they are based on a thorough understanding of how people process information. In the current paper we test this assumption in the context of social tagging systems. Cognitive research on how people assign tags has shown that they draw on two interconnected levels of knowledge in their memory: on a conceptual level of semantic fields or LDA topics, and on a lexical level that turns patterns on the semantic level into words. Another strand of tagging research reveals a strong impact of time-dependent forgetting on users' tag choices, such that recently used tags have a higher probability being reused than "older" tags. In this paper, we align both strands by implementing a computational theory of human memory that integrates the two-level conception and the process of forgetting in form of a tag recommender. Furthermore, we test the approach in three large-scale social tagging datasets that are drawn from BibSonomy, CiteULike and Flickr.

As expected, our results reveal a selective effect of time: forgetting is much more pronounced on the lexical level of tags. Second, an extensive evaluation based on this observation shows that a tag recommender interconnecting the semantic and lexical level based on a theory of human categorization and integrating time-dependent forgetting on the lexical level results in high accuracy predictions and outperforms other well-established algorithms, such as Collaborative Filtering, Pairwise Interaction Tensor Factorization, FolkRank and two alternative time-dependent approaches. We conclude that tag recommenders will benefit from going beyond the manifest level of word co-occurrences, and from including forgetting processes on the lexical level.

Keywords: Personalized tag recommendations · Time-dependent recommender systems · Latent Dirichlet Allocation · LDA · Human categorization · Human memory model · BibSonomy · CiteULike · Flickr

© Springer International Publishing Switzerland 2015
M. Atzmueller et al. (Eds.): MUSE/MSM 2013, LNAI 8940, pp. 75–95, 2015.
DOI: 10.1007/978-3-319-14723-9_5

1 Introduction

Many interactive systems are designed to mimic human behavior and thinking. A telling example for this are intelligent tutoring systems, which make inferences similar to teachers when drawing on knowledge of learning domains, knowledge about the learners and knowledge about effective teaching strategies. When looking at recommender systems, Collaborative Filtering approaches use information about socially related individuals to recommend items, much in the same way as humans are influenced by related peers when making choices. An implicit assumption behind this may be that interactive systems will perform better the closer they correspond to human behavior. This assumption seems to be reasonable as it is humans that interact with these systems, while these systems often also draw on data produced by humans (e.g., in the case of Collaborative Filtering). Therefore it can be assumed, that strategies that have evolved in humans over their individual or collective development form good models for interactive systems. However, the assumption that an interactive system will perform better the closer it mimics human behavior has not often been tested directly.

In the current paper, we investigate this assumption in the context of a tag recommender algorithm that borrows its basic architecture from MINERVA2 ([1], see also [2]), a computational theory of human categorization. We draw on research that has explored how human memory is used in a dynamic and adaptive fashion to understand new information encountered in the environment. Sensemaking happens by dynamically forming ad-hoc categories that relate the new information with knowledge stored in the semantic memory (e.g., [3]). For instance, when reading an article about "personalized recommendations", a novice has to figure out meaningful connections between previously distinct topics such as "cognition" and "information retrieval" and hence, has to start developing an ad-hoc category about common features of both of them. When using a social tagging system in such a situation, people apply labels to their own resources which to some extent externalizes this process of spontaneously generating ad-hoc categories [4]. Usually, a user describes a particular bookmark by a combination of about three to five tags verbalizing and associating aspects of different topics (e.g., "memory","retrieval", "recommendations", "collaborative filtering").

In previous work, we have shown that this behavior can be well described by differentiating between two separate forms of information processing. In human memory we find a semantic process that generates and retrieves topics or gist traces, and a verbal process that generates verbatim word forms to describe the topics [5]. In this paper we improve this model emphasizing on another fundamental principle of human cognition. According to Polyn et al. [6], memory traces including recently activated features contribute more strongly to retrieval than traces including features that have not been activated for a longer period of time. This relationship provides a natural account of what is called the recency effect in memory psychology (e.g., [7]). Obviously, things that happened a longer time ago tend to be forgotten and influence our current behavior less than things that have happened recently.

The purpose of this paper is twofold. First, we study the interaction between the effect of recency and the level of knowledge representation in human memory (semantic vs. verbal) within a social tagging system. In particular, we raise the question whether the impact of recency interacts with the level of knowledge representation, i.e., whether a time-dependent shift in the use of topics can be dissociated from a time-dependent shift in the use of particular tags. The second aim is to investigate to which extend our tag recommender based on MINERVA2 can be improved by integrating a time-dependent forgetting process. We also determine the performance of this recommender compared to other well-established tag recommender algorithms (e.g., Collaborative Filtering, FolkRank and Pairwise Interaction Tensor Factorization), as well as two alternative time-dependent approaches called GIRPTM [8] and BLL+C [9] (based on the ACT-R theory of human memory [10]). Hence, we raise the following two research questions:

- *RQ1*: Is there a difference between the time-dependent shift in the use of topics and the time-dependent shift in the use of particular tags?
- *RQ2*: Can a time-dependent forgetting process be integrated into a tag-recommender to create an efficient algorithm in comparison to the state-of-the-art?

The remainder of this paper tackles this two research questions and is organized as follows: We begin with discussing related work in the field of tag recommender in Sect. 2. Next, we review some of the work concerning recency in memory research and its current use in social tagging in Sect. 3 (first research question). Then we describe our approach and the experimental setup of our extensive evaluation in Sects. 4 and 5. Section 6 presents the results of this evaluation in terms of recommender quality (second research question). We conclude the paper by discussing our findings and future work in Sect. 7.

2   Related Work

Tagging as an important feature of the Social Web, has demonstrated to improve search considerably [11,12] and has supported the users with a simple tool to collaboratively organize and annotate content [13]. However, despite the potential advantages of tag usage, people do not tend to provide tags thoroughly or regularly. Thus, from an applied perspective, one important purpose of tag recommendations is to increase user's motivation to provide appropriate tags to their bookmarked resources.

In contrast to previously developed and typically data-driven tag recommender approaches, our research explores the suitability of psychologically sound memory processes to improve tag recommender approaches. Previously, in [5,9] we presented two simple methods (= 3L and BLL+C) that aim to explain memory processes in social tagging systems. Based on our previous research and other incentives from related work, we introduce in this work a novel time-based tag recommender algorithm (= $3LT_{tag}$) based on the MINERVA2 theory of human

categorization [1,2] that significantly outperforms popular state-of-the-art algorithms as well as BLL+C [9], an alternative time-based approach based on the
ACT-R theory of human memory [10]. It models the activation of elements in
a person's declarative memory by considering frequency and recency of a user's
tagging history as well as semantic context.

To date, two tag-recommender approaches have been established: graph-
based and content-based tag recommender systems [14], whereas in this work
we focus on graph-based approaches. Prominent algorithms in this respect can
be found for instance in the work of Hotho et al. [15] who introduced FolkRank
(FR), which has established itself as the most prominent benchmarking tag recommender approach over the past few years. Further investigated, was the recommendation of tags to users in a personalized manner. In the scope of this
research strand, Jäschke et al. [16] or Hamouda and Wanas [17] are well known
to present a set of Collaborative Filtering (CF) approaches. Rendle et al. [18],
Krestel et al. [19] or Rawashdeh et al. [20] more recently presented a factorization
model (FM and PITF), a semantic model (based on LDA) or a link prediction
model to recommend tags to users, respectively (see also Sect. 5.3).

Comparing these principles now with simple "most popular tags" approaches,
we will notice a big disadvantage in their computational expense as well as in
their lack of considering recent observations made in social tagging systems, such
as the variation of the individual tagging behavior over time [21]. To that end,
recent research has made first promising steps towards more accurate graph-
based models that also account for the variable of time [8,22].

However, although these time-dependent approaches have shown to outperform some of the current state-of-the-art tag recommender algorithms, all of
them ignore well-established and long standing research from cognitive psychology on how humans process information. Therefore, we try to fill this gap by
investigating tagging mechanisms that aim to mimic peoples' tagging behavior.

3   Recency in Memory and in the Use of Social Tagging

In previous work we have introduced 3Layers [5], which is a model for recommending tags that is inspired by cognitive-psychological research on categorizing and
verbalizing objects (e.g., [4]) and is adapted in this work based on MINERVA2 in
order to answer our two research questions. 3Layers consists of an input, a hidden
and an output layer, where the hidden layer is built up by a semantic and an interconnected lexical matrix. The semantic matrix stores the topics of all bookmarks
in the user's personomy[1], calculated with Latent Dirichlet Allocation (LDA) [19],
while the lexical matrix stores the tags of those bookmarks. In a first step of calculation, the LDA topics of a new bookmark, for which appropriate tags should be
recommended, are represented in the input layer and compared with the semantic
matrix of the hidden layer. In the course of this comparison, semantically relevant

---

[1] We define a bookmark (also known as "post") as the set of tags a target user has
assigned to a target resource at a specific time, and the personomy as a collection of
all bookmarks of a user.

bookmarks of the user's personomy become activated. The resulting pattern of activation across the semantic matrix is then applied to the lexical matrix to further activate and recommend those tags that belong to relevant bookmarks. In a final step, the activation pattern across the lexical matrix is summarized on the output layer in form of a vector. This vector represents a tag distribution that can be used to predict a substantial amount of variance in the user's tagging behavior when creating a new bookmark.

We draw on Fuzzy Trace Theory (FTT; e.g., [23]) to make a prediction with respect to our first research question about a potentially differential impact of recency on semantic and lexical representations, i.e., on the usage of topics and tags, respectively. FTT differentiates between two distinct memory traces, a gist trace and a verbatim trace, which represent general semantic information of e.g., a read sentence and the sentence's exact wording, respectively. These two types of memory traces share properties with our distinction between a semantic and a lexical matrix (see also Sect. 4). While vectors of the semantic matrix provide a formal account of each bookmark's gist (its general semantic content), vectors of the lexical matrix correspond to a bookmark's verbatim trace (explicit verbal information in form of assigned tags). This distinction is also in line with Kintsch and Mangalath [24] who model gist traces of words by means of LDA topic vectors and explicit traces of words by means of word co-occurrence vectors. An empirically well-established assumption of FTT is that verbatim traces are much more prone to time-dependent forgetting than gist traces (e.g., [23]): while people tend to forget the exact wording, usually they can remember the gist of a sentence (or a bookmark). Taken together, we derived the hypothesis that a user's verbatim traces (vectors in the lexical matrix that encode the user's tags) are more strongly affected by time-dependent forgetting and therefore more variable over time than a user's gist traces (vectors in the semantic matrix that contain topics).

To test this hypothesis, we performed an empirical analysis in BibSonomy, CiteULike and Flickr (see Sect. 5.1). The topics for the resources of these datasets' bookmarks were calculated using Latent Dirichlet Allocation (LDA) [19] (see Sect. 4.2) based on 100, 500 and 1000 latent topics in order to cover different levels of topic specialization (these numbers of latent topics are also suggested by related work in the field [24,25]). For each user we selected the most recent bookmark (i.e., the one from the test set with the most recent timestamp, see also Sect. 5.2) and described the bookmark by means of two vectors: one encoding the bookmark's LDA topic pattern (gist vector) and one encoding the tags assigned by the user (verbatim vector). Then, we searched for all the remaining bookmarks of the same user, described each of them by means of the two vectors and arranged them in a chronologically descending order. Next, we compared the gist and the verbatim vector of the most recent bookmark with the two corresponding vectors of all bookmarks in the user's past by means of the cosine similarity measure.

The obtained results are represented in the three diagrams of Fig. 1, plotting the average cosine similarities over all users against the time lags (given in

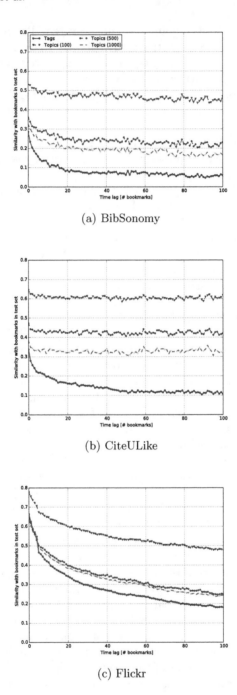

(a) BibSonomy

(b) CiteULike

(c) Flickr

Fig. 1. Interaction between time-dependent forgetting and level of knowledge representation for BibSonomy, CiteULike and Flickr showing a more pronounced decline for tags than for topics (100, 500, 1000 LDA topics; first research question).

number of past bookmarks). For all three datasets we show these results for the last 100 bookmarks of tagging activity per user because in this range, there are enough users available for each past bookmark to calculate mean values reliably. The diagrams quite clearly reveal that – independent of the environment (Bib-Sonomy, CiteULike or Flickr) – the similarity between the most recent bookmark and all other bookmarks decreases monotonically as a function of time lag. More importantly and as expected, the time-dependent decline is more strongly pronounced for the verbatim vectors (encoding tag assignments) in contrast to the gist vectors (encoding LDA topics). Furthermore, we can see that the more LDA topics we use, the more similar is the time-dependent decline of the two vectors (tags vs. topics) to each other.

4   Approach

In this section we introduce two novel time-dependent tag recommender algorithms which model the process of forgetting on a semantic and lexical layer in a time-depended manner. Moreover, we describe how we created the semantic features (i.e., topics) for the bookmarks in our datasets using *Latent Dirichlet Allocation* (LDA).

4.1   Tag-Recommender Algorithms

Due to our findings introduced within the previous section, we assume that the factor of time plays a more critical role on the lexical layer than on the semantic layer. The approaches implemented in this section are based on a preliminary recommender model called 3Layers (3L) that was introduced in our previous work [5].

   Figure 2 schematically shows how 3Layers (3L) represents a user's personomy within the hidden layer, which interconnects a semantic matrix, M_S (l bookmarks $\cdot$ n LDA topics matrix), and a lexical matrix, M_L (l bookmarks $\cdot$ m tags matrix). Thus, each bookmark of the user is represented by two associated vectors; by a vector of LDA topics $S_{i,k}$ stored in M_S and by a vector of tags $L_{i,j}$ stored in M_L. Similar to [2], we borrow a mechanism from MINERVA2, a computational theory of human categorization [1], to process the network constituted by the input, hidden and output layer. First, the LDA topics of the target resource to be tagged are represented on the input layer in form of a vector P with n features. Then, P is used as a cue to activate each bookmark (B_i) in M_S depending on the similarity (Sim_i) between both vectors, i.e., P and B_i. Similar to [2], we estimate Sim_i by calculating the cosine between the two vectors:

$$Sim_i = \frac{\sum_{k=1}^{n}(P_k \cdot S_{i,k})}{\sqrt{\sum_{k=1}^{n} P_k^2} \cdot \sqrt{\sum_{k=1}^{n} S_{i,k}^2}} \tag{1}$$

If no topics are available for the target resource (i.e., $n = 0$), we set Sim_i to 1 and thus, activate each bookmark with the same value. To transform the

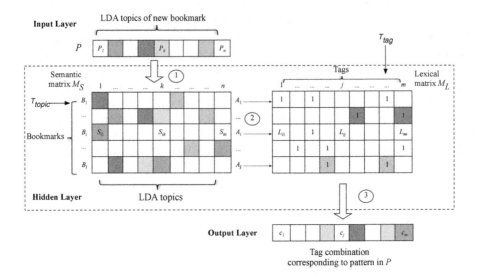

Fig. 2. Schematic illustration of 3L showing the connections between the semantic matrix (M_S) encoding the LDA topics and the lexical matrix (M_L) encoding the tags. Furthermore, T_{topic} and T_{tag} schematically demonstrate how the time component is integrated in case of $3LT_{topic}$ and $3LT_{tag}$, respectively.

resulting similarity values into activation values (A_i) and to further reduce the influence of bookmarks with low similarities, Sim_i is raised to the power of 3, i.e., $A_i = Sim_i^3$ (see also [1]). Next, these activation values are propagated to M_L to activate tags that are associated with highly activated bookmarks on the semantic matrix M_S (circled numbers 2 and 3 in Fig. 2). This is computed by the following equation that yields an activation value c_j for each of the m tags on the output layer:

$$c_j = \underbrace{\sum_{i=1}^{l}(L_{i,j} \cdot A_i)}_{3L} \qquad (2)$$

To finally compute $3LT_{topic}$ and $3LT_{tag}$, we integrate a time component on the level of topics (hereinafter called T_{topic}) and on the level of tags (T_{tag}), respectively. Both recency components are calculated by the following equation that is based on the base-level learning (BLL) equation [7]:

$$BLL(t) = ln((tmstp_{ref} - tmstp_t)^{-d}), \qquad (3)$$

where $tmstp_{ref}$ is the timestamp of the most recent bookmark of the user and $tmstp_t$ is the timestamp of the last occurrence of t, encoded as the topic in the case of T_{topic} or as the tag in the case of T_{tag}, in the user's bookmarks. The exponent d accounts for the power-law of forgetting and was set to .5 as suggested by Anderson et al. [10]. While $3LT_{topic}$ can be computed by using Eq. 4, $3LT_{tag}$ can be computed by using Eq. 5:

$$c_j = \sum_{i=1}^{l}(L_{i,j} \cdot \underbrace{\sum_{k=1}^{n}(S_{i,k} \cdot BLL(k)) \cdot A_i)}_{3LT_{topic}} \tag{4}$$

$$c_j = \underbrace{\sum_{i=1}^{l}(L_{i,j} \cdot BLL(j) \cdot A_i)}_{3LT_{tag}} \tag{5}$$

As suggested in related work [9,14,26], we additionally consider tags that have been applied to the target resource by other users. This allows the recommendation of new tags, i.e., tags that have not been used by the target user before. We implement this by taking into account the most popular tags in the tag assignments of the resource Y_r (MP_r, i.e., $\arg\max_{t \in T}(|Y_r|)$) [15]. Therefore, we have chosen MP_r over other methods like CF, as previous work [27,28] shows that users in social tagging systems are more likely to imitate previously assigned tags by other users to a target resource. In order to combine c_j with MP_r, the following normalization method was used:

$$\|c_j\| = \frac{exp(c_j)}{\sum_{i=1}^{m} exp(c_i)} \tag{6}$$

Taken together, the list of recommended tags for a given user u and resource r is then calculated as

$$\tilde{T}(u,r) = \arg\max_{j \in T}(\beta\|c_j\| + (1-\beta)\||Y_r|\|), \tag{7}$$

where β is used to inversely weight the two components. The results presented in Sect. 6 were calculated using $\beta = .5$, thus, applying the same weight to both components.

4.2   Topic Generation via LDA

As outlined in Sect. 3, we used LDA to calculate the semantic features (i.e., topics) of the resources of the full datasets. LDA is a probability model that helps to find latent topics for documents where each topic is described by words in these documents [19]. This can be formalized as follows:

$$P(t_i|d) = \sum_{j=1}^{Z}(P(t_i|z_i = j) \cdot P(z_i = j|d)) \tag{8}$$

Table 1. Properties of the used dataset samples, where $|B|$ is the number of bookmarks, $|U|$ the number of users, $|R|$ the number of resources, $|T|$ the number of tags and $|TAS|$ the number of tag assignments.

Dataset	$|B|$	$|U|$	$|R|$	$|T|$	$|TAS|$
BibSonomy	400,983	5,488	346,444	103,503	1,479,970
CiteULike	379,068	8,322	352,343	138,091	1,751,347
Flickr	864,679	9,590	864,679	127,599	3,552,540

Here $P(t_i|d)$ is the probability of the ith word for a document d and $P(t_i|z_i = j)$ is the probability of t_i within the topic z_i. $P(z_i = j|d)$ is the probability of using a word from topic z_i in the document. The number of latent topics Z is determined in advance and defines the level of granularity. We calculated the semantic features for our datasets with different amounts of LDA topics (100, 500 and 1000 - see also [24,25]).

When using LDA in tagging environments, documents are resources which are described by tags. This means that based on the tag vectors of the resources (i.e., all the tags the users have assigned to the resource), resources in the bookmarks can also be represented with the topics identified by LDA. These topics were then used as features in the semantic matrix M_S. We implemented LDA with Gibbs sampling using the Java framework Mallet[2].

5   Experimental Setup

In this section we describe our experiment's datasets, evaluation methodology and the baseline algorithms in detail.

5.1   Datasets

To conduct our study, we used three well-known folksonomy datasets that are freely available for scientific purposes and thus, allow for reproducibility. In this respect, we utilized datasets from the social bookmark and publication sharing system BibSonomy[3] (2013-07-01), the reference management system CiteULike[4] (2013-03-10) and the image sharing platform Flickr[5] (2010-01-07) to evaluate our approach on both types of folksonomies, broad (BibSonomy and CiteULike; all users are allowed to annotate a particular resource) and narrow (Flickr; only the user who has uploaded a resource is allowed to tag it) ones [29]. We furthermore excluded all automatically generated tags from the datasets (e.g., *no-tag*, *bibtex-import*, etc.) and decapitalized all tags as suggested in related work (e.g., [18]).

---

[2] http://mallet.cs.umass.edu/topics.php.
[3] http://www.kde.cs.uni-kassel.de/bibsonomy/dumps/.
[4] http://www.citeulike.org/faq/data.adp.
[5] http://www.tagora-project.eu/data/#flickrphotos.

To reduce computational effort, we randomly selected 10 % of CiteULike, and 3 % of Flickr user profiles (see also [30])[6]. We did not apply a p-core pruning to keep the original bookmarks of the users and thus, to prevent a biased evaluation [31]. The statistics of our used dataset samples can be found in Table 1.

5.2   Evaluation Methodology

To evaluate our tag recommender approaches, we split the three datasets into training and test sets based on a leave-one-out hold-out method as proposed in related work (e.g., [16]). Hence, for each user we selected her most recent bookmark (or post) in time and put it into the test set. The remaining bookmarks were then used for the training of the algorithms. This procedure is a promising simulation of a real-world environment, as it predicts a user's future tagging behavior based on tagging behavior in the past. Furthermore, it is a standard practice for evaluation of time-based recommender systems [32].

In order to quantify the recommender quality and to benchmark our recommender against other tag recommendation approaches, a set of well-known metrics in information retrieval and recommender systems were used [14, 16]:

Recall (R) is defined as the number of recommended tags that are relevant for the target user/resource divided by the total number of relevant tags [33]:

$$R@k = \frac{1}{|U|} \sum_{u \in U} \left(\frac{|t_u^k \cap T_u|}{|T_u|} \right),$$ (9)

where t_u^k denotes the top k recommended tags and T_u the list of relevant tags of a bookmark of user $u \in U$.

Precision (P) is calculated as the number of correctly recommended tags divided by the total number of recommended tags $|t_u^k|$ $(= k)$ [33]:

$$P@k = \frac{1}{|U|} \sum_{u \in U} \left(\frac{|t_u^k \cap T_u|}{|t_u^k|} \right)$$ (10)

F1-score (F1) is a combination of the recall and precision metrics and is calculated using the following equation [33]:

$$F1@k = \frac{1}{|U|} \sum_{u \in U} \left(2 \cdot \frac{P@k \cdot R@k}{P@k + R@k} \right)$$ (11)

Mean reciprocal rank (MRR) is a rank-dependent evaluation metric that is calculated as the sum of the reciprocal ranks (or positions) of all relevant tags in the list of recommended tags [20]:

$$MRR = \frac{1}{|U|} \sum_{u=1}^{|U|} \left(\frac{1}{|T_u|} \sum_{t \in T_u} \frac{1}{rank(t)} \right)$$ (12)

---

[6] **Note:** We used the same dataset samples as in our previous work [9], except for CiteULike, where we used a smaller sample for reasons of computational effort in respect to the calculation of the LDA topics.

This way, a recommender achieves a higher MRR if relevant tags occur at early positions in the list of recommended tags.

Mean average precision (MAP) extends the precision metric and also considers the order of the recommended tags. This is done by computing the precision value at every position k of the ranked list of tags and using the average of these values [20]:

$$MAP = \frac{1}{|U|} \sum_{u=1}^{|U|} \left(\frac{1}{|T_u|} \sum_{k=1}^{|t_u^k|} (B_k \cdot P@k) \right), \tag{13}$$

where B_k is 1 if the tag at position k of the list of recommended tag is correct.

In particular, we report R@k, P@k, MRR and MAP for $k = 10$ and F1-Score ($F_1@k$) for $k = 5$ recommended tags.

5.3   Baseline Algorithms

We compared the results of our approach to several "baseline" tag recommender algorithms. The algorithms were selected in respect to their popularity in the community, performance and novelty [34]. The most basic approach we utilized is the unpersonalized *MostPopular (MP)* algorithm. MP recommends independent of user and resource, the same set of tags that is weighted by the frequency over all tag assignments [16]. A personalized extension of MP is the *MostPopular$_{u,r}$* ($MP_{u,r}$) algorithm that suggests the most frequent tags in the tag assignments of the user (MP_u) and the resource (MP_r) [16]. As done in our approaches, we weighted the user and the resource components equally ($\beta = .5$).

Another well known recommender approach is *Collaborative Filtering (CF)* which was adapted for tag recommendations by Marinho et al. [34]. Here the neighborhood of a user is formed based on the tag assignments in the user profile and the only variable parameter is the number of users k in this neighborhood. k has been set to 20 as suggested by Gemmell et al. [30]. In Sect. 4.2 we have described how we applied *Latent Dirichlet Allocation (LDA)* for tag recommendations. The results presented in this work have been calculated using $Z = 1000$ latent topics [19].

An additional approach we utilized is the well-known *FolkRank (FR)* algorithm which is an improvement of the *Adapted PageRank (APR)* approach [16]. FR extends the PageRank algorithm in order to rank the nodes within the graph structure of a folksonomy [16], which is based on their importance in the network. Our implementation of APR and FR builds upon the code and the settings of the open-source Java tag recommender framework provided by the University of Kassel[7]. In this implementation the parameter d is set to .7 and the maximum number of iterations l is set to 10.

A different popular and recent tag recommender mechanism is *Pairwise Interaction Tensor Factorization (PITF)* proposed by Rendle and Schmidt-Thieme [18]. It is an extension of *Factorization Machines (FM)* and explicitly models pairwise interactions between users, resources and tags. The FM

---

[7] http://www.kde.cs.uni-kassel.de/code.

and PITF results presented in this paper were calculated using the open-source C++ tag recommender framework provided by the University of Konstanz[8]. We set the dimensions of factorization k_U, k_R and k_T to 256 and the number of iterations l to 50 as suggested in [18].

Finally, we tried to benchmark against two time-dependent approaches. The first one is the *GIRPTM* algorithm presented by Zhang et al. [8] which is based on the frequency and the temporal usage of a user's tag assignments. The approach models the temporal tag usage with an exponential distribution based on the first- and last-time usage of the tags. The second time-dependent tag-recommender approach is the *Base-Level Learning Equation with Context (BLL+C)* algorithm introduced in our previous work [9]. BLL+C is based on the ACT-R human memory theory by Anderson et al. [10] and uses a power-law distribution based on all tag usages to mimic the time-dependent forgetting in tag applications. In both approaches the resource component is modeled by a simple most popular tags by resource mechanism, as it is also done in our 3Layers approach. In previous work [9], we showed that BLL+C outperforms GIRPTM and other well-established algorithms, such as FR, PITF and CF.

The algorithms described in this section along with our developed approaches (see Sect. 4 are implemented within our Java-based *TagRec* framework [35]. Published as open-source software, it can be downloaded from our Github Repository[9] along with the herein used test and training sets (see Sects. 5.1 and 5.2).

6   Results

In this section we present the evaluation of the two novel algorithms in line with our research questions. In step 1, we compared the three 3Layers approaches (3L, $3LT_{topic}$ and $3LT_{tag}$) with one another, in order to examine our first research question of whether recency has a differential effect on topics and tags. According to the empirical analysis illustrated in Sect. 3, $3LT_{tag}$ yields more accurate predictions than $3LT_{topic}$ and 3L.

Results shown in Table 2 prove this assumption since - independent of the metric (F_1@5, MRR and MAP) and the number of LDA topics (100, 500, and 1000) applied - the difference between $3LT_{tag}$ and 3L is significantly larger than the one between $3LT_{topic}$ and 3L. This allows us to conclude that a user's gist traces (LDA topics) associated with the user's bookmarks are less prone to "forgetting" than a user's verbatim traces (tags associated with the bookmarks). Interestingly, this effect is more strongly pronounced under the narrow folksonomy condition (Flickr), where no tags of other users are available for the target user's resource, than under the broad folksonomy condition (BibSonomy and CiteULike), where users could get inspired by tags of other users.

Furthermore, Table 2 illustrates the performance of 3L, $3LT_{topic}$ and $3LT_{tag}$ for different numbers of LDA topics (100, 500 and 1000). It can be seen that

---

[8] http://www.informatik.uni-konstanz.de/rendle/software/tag-recommender/.

[9] https://github.com/learning-layers/TagRec/.

Table 2. $F_1@5$, MRR and MAP values for BibSonomy, CiteULike and Flickr showing the performance of 3L and its time-dependent extensions ($3LT_{topic}$ and $3LT_{tag}$) for 100, 500 and 1000 LDA topics (first research question).

	# Topics	Measure	3L	$3LT_{topic}$	$3LT_{tag}$
BibSonomy	100	$F_1@5$	.197	.198	**.204**
		MRR	.152	.154	**.161**
		MAP	.201	.202	**.212**
	500	$F_1@5$	.204	.205	**.209**
		MRR	.156	.158	**.163**
		MAP	.206	.208	**.215**
	1000	$F_1@5$	.206	.207	**.211**
		MRR	.157	.158	**.162**
		MAP	.207	.208	**.214**
CiteULike	100	$F_1@5$	.211	.212	**.221**
		MRR	.192	.194	**.211**
		MAP	.226	.228	**.248**
	500	$F_1@5$	.218	.219	**.225**
		MRR	.196	.198	**.211**
		MAP	.232	.234	**.250**
	1000	$F_1@5$	.232	.233	**.238**
		MRR	.199	.200	**.212**
		MAP	.235	.236	**.250**
Flickr	100	$F_1@5$	.500	.507	**.535**
		MRR	.421	.429	**.476**
		MAP	.560	.571	**.634**
	500	$F_1@5$	.564	.567	**.582**
		MRR	.443	.448	**.476**
		MAP	.591	.596	**.635**
	1000	$F_1@5$	.568	.571	**.585**
		MRR	.450	.454	**.477**
		MAP	.599	.604	**.636**

all three approaches provide good results for different levels of topic specialization, with the best accuracy values reached for 1000 LDA topics[10]. $F_1@5$, MRR and MAP values calculated for 1000 topics are further used within the second evaluation step, which is described in the next paragraph.

In a second step, we compared the performance of our approaches, especially $3LT_{tag}$, with several state-of-the-art algorithms. By this means we address our second research question, of whether 3L and its two extensions can be implemented in form of effective and efficient tag recommendation mechanisms. First, Table 3 reveals that all personalized recommendation mechanisms clearly outperform the unpersonalized MP approach. This is not surprising, as MP solely takes into account the tag's usage frequency independent of information about a particular user or resource.

Second and more important, 3L and its two extensions ($3LT_{topic}$ and $3LT_{tag}$) reach significantly higher accuracy estimates than the well-established mechanisms LDA, $MP_{u,r}$, CF, APR, FR, FM and PITF. From this we conclude that

---

[10] **NOTE:** We also performed experiments with more than 1000 LDA topics (e.g., 2000, 3000, ...). However, as also shown by related work (e.g., [19,24,25]) this step did not help in increasing the performance of the LDA-based tag recommenders.

Table 3. $F_1@5$, MRR and MAP values for all the users in the datasets (BibSonomy, CiteULike and Flickr) and for users with a minimum number of 20 bookmarks ($B_{min} = 20$) showing that our time-dependent $3LT_{tag}$ approach outperforms current state-of-the art algorithms (second research question). The symbols *, ** and *** indicate statistically significant differences based on a Wilcoxon Ranked Sum test between 3L, $3LT_{topic}$, $3LT_{tag}$ and BLL+C at α level .05, .01 and .001, respectively; °, °° and °°° indicate statistically significant differences between our two time-dependent approaches $3LT_{topic}$, $3LT_{tag}$ and 3L at the same α levels.

Dataset	B_{min}	Measure	MP	LDA	MP_u	MP_r	$MP_{u,r}$	CF	APR	FR	FM	PITF	GIRPTM	BLL+C	3L	$3LT_{topic}$	$3LT_{tag}$
BibSonomy	-	$F_1@5$	.013	.097	.152	.074	.192	.166	.175	.171	.122	.139	.197	.201	.206	.207	**.211**
		MRR	.008	.083	.114	.054	.148	.133	.149	.148	.097	.120	.152	.158	.157	.158	**.162**
		MAP	.009	.101	.148	.070	.194	.173	.193	.194	.120	.150	.200	.207	.207	.208	**.214**
	20	$F_1@5$	.019	.142	.156	.078	.195	.204	.184	.197	.162	.163	.240	.249	.264	.269	**.296**°**
		MRR	.011	.129	.135	.059	.160	.175	.159	.171	.135	.137	.201	.216	.224	.227	**.251****
		MAP	.012	.152	.163	.074	.200	.219	.197	.214	.164	.166	.256	.275	.289	.291	**.325°**
CiteULike	-	$F_1@5$	.007	.068	.182	.033	.199	.157	.162	.160	.113	.130	.207	.215	.232	.233	**.238****
		MRR	.005	.065	.164	.024	.179	.168	.181	.181	.116	.149	.196	.205	.199	.200	**.212**
		MAP	.005	.073	.191	.029	.210	.196	.212	.212	.132	.169	.229	.241	.235	.236	**.250**
	20	$F_1@5$	.008	.145	.228	.031	.237	.228	.221	.225	.193	.196	.282	.298	.331*	.334*	**.353*****
		MRR	.006	.144	.225	.022	.233	.271	.237	.239	.201	.210	.321	.335	.312	.316	**.367°°°**
		MAP	.006	.162	.258	.028	.269	.308	.273	.276	.229	.237	.369	.389	.369	.373	**.430°°°**
Flickr	-	$F_1@5$	.023	.169	.435	-	.435	.417	.328	.334	.297	.316	.509	.523	.568***	.571***	**.585*****
		MRR	.023	.171	.360	-	.360	.436	.352	.355	.300	.333	.445	.466	.450	.454	**.477°°°**
		MAP	.023	.205	.468	-	.468	.581	.453	.459	.384	.426	.590	.619	.599	.604	**.636°°°**
	20	$F_1@5$	.030	.190	.382	-	.382	.495	.322	.334	.309	.309	.534	.553	.610***	.616***	**.643°°°**
		MRR	.028	.174	.322	-	.322	.473	.309	.317	.290	.289	.485	.508	.478	.485	**.530****
		MAP	.029	.215	.427	-	.427	.655	.405	.419	.378	.376	.664	.701	.661	.670	**.732°°°**

predicting tags based on psychologically plausible steps that turn a user's gist traces into words, calculates tag recommendations that correspond well to the user's tagging behavior.

Third, we can see that also the two other time-dependent algorithms (GIRPTM and BLL+C) outperform the state-of-the art approaches that do not take the time component into account. BLL+C based on ACT-R even reaches slightly higher estimates of accuracy than our 3L approach based on MINERVA2. However, this relation changes when we enhance 3L by the recency component at the level of tags. Then, $3LT_{tag}$ clearly outperforms BLL+C with respect to all three metrics and across all three datasets. Finally, as shown in Fig. 3, a very similar pattern of results becomes apparent when evaluating the different approaches by plotting recall against precision for $k = 1$–10 recommended tags.

To furthermore prove our assumption that memory processes play an important role in social tagging systems, we also performed an experiment where we looked at users that have bookmarked a minimum of $B_{min} = 20$ resources (see also [36]). We conducted this experiment by applying a post-filtering method, i.e., recommendations were still calculated on the whole folksonomy graph but accuracy estimates were calculated only on the basis of the filtered user profiles (= 780 users in the case of BibSonomy, 1,757 in the case of CiteULike and 4,420 for Flickr). The results of the experiment are also shown in Table 3. We can observe that in general the accuracy estimates of all algorithms are increasing. Furthermore, it demonstrates that the difference between $3LT_{tag}$ and the other algorithms (including BLL+C) grows substantially larger the more user "memory" (history) is used. These differences between $3LT_{tag}$ and BLL+C as well as between $3LT_{tag}$ and 3L proved to be statistically significant based on a Wilcoxon Rank Sum test across all accuracy metrics (F_1@5, MRR and MAP) and all three datasets (see Table 3).

7   Discussion and Conclusion

In this study we have provided empirical evidence for an interaction between the level of knowledge representation (semantic vs. lexical) and time-based forgetting in the context of social tagging. Based on the analysis of three large-scale tagging datasets (BibSonomy, CiteULike and Flickr), we conclude that - as expected - the gist traces of a user's personomy (the combination of LDA topics associated with the bookmarks) are more stable over time than the verbatim traces (the combination of associated tags). This pattern of results is well in accordance with research on human memory (e.g., [23]) suggesting that while people tend to forget surface details they keep quite robust memory traces of the general meaning underlying the experiences of the past (e.g., the meaning of read words). The interaction effect suggests that it is worthwhile to differentiate between both, time-based forgetting as well as the level of knowledge representation in social tagging research. Moreover, the differential affect of forgetting on the two levels of processing has further substantiated the differences between tagging behavior on a semantic level of gist traces and a lexical level of verbatim traces [28].

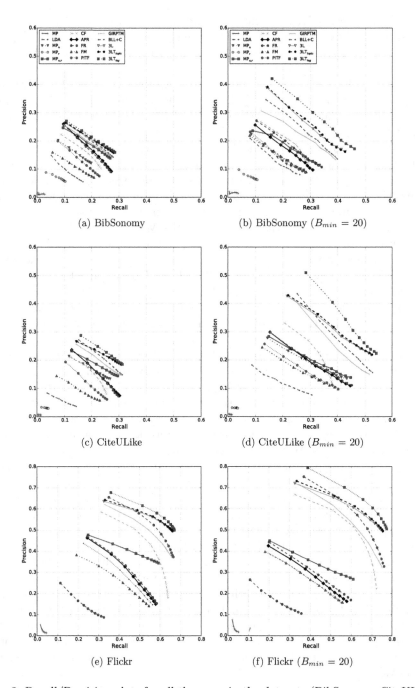

Fig. 3. Recall/Precision plots for all the users in the datasets (BibSonomy, CiteULike and Flickr) and for users with a minimum number of 20 bookmarks ($B_{min} = 20$) showing the performance of the algorithms for 1–10 recommended tags (k).

This in turn is in line with cognitive research on social tagging (e.g., [37]) that suggests to consider a latent, semantic level (e.g., modeled in form of LDA topics) when trying to understand the variance in the statistical patterns on the manifest level of users' tagging behavior.

Finally, we have gathered evidence for our assumption that interactive systems can be improved by basing them on a thorough understanding of how humans process information. We note in particular that integrating two fundamental principles of human information processing, time-based forgetting and differentiating into semantic and lexical processing, enhances the accuracy of tag predictions as compared to a situation when only one of the principles is considered. Our experiments showed that topics are more stable over time which means that they are, unless tags, not as suitable to be modelled using the BLL equation but can improve the results as an activation value on the basis of topic similarities. Therefore, 3L, that is based on the MINERVA2 theory of human categorization [1, 2] is enhanced by forgetting on the lexical level ($3LT_{tag}$). This approach significantly outperforms both the traditional 3L, as well as other well-established algorithms, such as CF, APR, FR, FM, PITF and the time-based GIRPTM. Furthermore, $3LT_{tag}$ also clearly reaches higher levels of accuracy than BLL+C, the to-date leading time-based tag recommender approach, that is based on the ACT-R theory of human memory [10] and was introduced in our previous work [9].

One limitation of this work is the calculation of semantic features (or topics) of the resources using LDA, which is not only very time-consuming but also could be biased because of the tag information it is based on. In this respect an interesting extension for future work would be to additionally conduct our experiments using external topics of the resources (e.g., Wikipedia categories as used in [5]). Looking at another aspect, our work has been inspired by the human memory model ACT-R proposed by Anderson et al. [10], but so far only investigates the first part of the equation, the recency component. Thus, it would be very interesting to further extend our approach by additionally investigating the associative component of the model. Also, as the computations were carried out with fixed values .5 for the exponent d (3) and the weight β (7), it would be worth exploring alternative values.

Moreover, we plan to include our algorithms in an actual online social tagging system (e.g., BibSonomy). Only in such a setting it is possible to test the recommendation performance by looking at user acceptance. Because our approach is theory-driven, it is rather straightforward to transfer it to recommendations in other interactive systems and Web paradigms where semantic and lexical processing plays a role (such as, for example, in Web curation). Thus, the generalization to other paradigms is another important benefit of driving recommender systems research by an understanding of human information processing on the Web.

Acknowledgments. This work is supported by the Know-Center, the EU funded projects Learning Layers (Grant Agreement 318209) and weSPOT (Grant Agreement

318499) and the Austrian Science Fund (FWF): P 25593-G22. Moreover, parts of this work were carried out during the tenure of an ERCIM "Alain Bensoussan" fellowship programme.

References

1. Hintzman, D.L.: Minerva 2: a simulation model of human memory. Behav. Res. Methods Instrum. Comput. **16**, 96–101 (1984)
2. Kwantes, P.J.: Using context to build semantics. Psychon. Bull. Rev. **12**, 703–710 (2005)
3. Barsalou, L.: Situated simulation in the human conceptual system. Lang. Cogn. Process. **18**, 513–562 (2003)
4. Glushko, R.J., Maglio, P.P., Matlock, T., Barsalou, L.W.: Categorization in the wild. Trends Cogn. Sci. **12**, 129–135 (2008)
5. Seitlinger, P., Kowald, D., Trattner, C., Ley, T.: Recommending tags with a model of human categorization. In: Proceedings of CIKM '13, pp. 2381–2386. ACM, New York (2013)
6. Polyn, S.M., Norman, K.A., Kahana, M.J.: A context maintenance and retrieval model of organizational processes in free recall. Psychol. Rev. **116**, 129 (2009)
7. Anderson, J.R., Schooler, L.J.: Reflections of the environment in memory. Psychol. Sci. **2**, 396–408 (1991)
8. Zhang, L., Tang, J., Zhang, M.: Integrating temporal usage pattern into personalized tag prediction. In: Sheng, Q.Z., Wang, G., Jensen, C.S., Xu, G. (eds.) APWeb 2012. LNCS, vol. 7235, pp. 354–365. Springer, Heidelberg (2012)
9. Kowald, D., Seitlinger, P., Trattner, C., Ley, T.: Long time no see: the probability of reusing tags as a function of frequency and recency. In: Proceedings of WWW '14. ACM, New York (2014)
10. Anderson, J.R., Byrne, M.D., Douglass, S., Lebiere, C., Qin, Y.: An integrated theory of the mind. Psychol. Rev. **111**, 1036–1050 (2004)
11. Helic, D., Trattner, C., Strohmaier, M., Andrews, K.: Are tag clouds useful for navigation? a network-theoretic analysis. Int. J. Soc. Comput. Cyber-Phys. Syst. **1**, 33–55 (2011)
12. Trattner, C., Lin, Y.l., Parra, D., Yue, Z., Real, W., Brusilovsky, P.: Evaluating tag-based information access in image collections. In: Proceedings of the 23rd ACM Conference on Hypertext and Social Media, pp. 113–122. ACM (2012)
13. Körner, C., Benz, D., Hotho, A., Strohmaier, M., Stumme, G.: Stop thinking, start tagging: tag semantics emerge from collaborative verbosity. In: Proceedings of the 19th International Conference on World Wide Web, WWW '10, pp. 521–530. ACM, New York (2010)
14. Lipczak, M.: Hybrid tag recommendation in collaborative tagging systems. Ph.D. thesis, Dalhousie University (2012)
15. Hotho, A., Jäschke, R., Schmitz, C., Stumme, G.: Information retrieval in folksonomies: search and ranking. In: Sure, Y., Domingue, J. (eds.) ESWC 2006. LNCS, vol. 4011, pp. 411–426. Springer, Heidelberg (2006)
16. Jäschke, R., Marinho, L., Hotho, A., Schmidt-Thieme, L., Stumme, G.: Tag recommendations in folksonomies. In: Kok, J.N., Koronacki, J., Lopez de Mantaras, R., Matwin, S., Mladenič, D., Skowron, A. (eds.) PKDD 2007. LNCS (LNAI), vol. 4702, pp. 506–514. Springer, Heidelberg (2007)
17. Hamouda, S., Wanas, N.: Put-tag: personalized user-centric tag recommendation for social bookmarking systems. Soc. Netw. Anal. Min. **1**, 377–385 (2011)

18. Rendle, S., Schmidt-Thieme, L.: Pairwise interaction tensor factorization for personalized tag recommendation. In: Proceedings of WSDM 2010, pp. 81–90. ACM, New York (2010)
19. Krestel, R., Fankhauser, P., Nejdl, W.: Latent dirichlet allocation for tag recommendation. In: Proceedings of RecSys 2009, pp. 61–68. ACM (2009)
20. Rawashdeh, M., Kim, H.N., Alja'am, J.M., El Saddik, A.: Folksonomy link prediction based on a tripartite graph for tag recommendation. J. Intell. Inf. Syst. 40(2), 307–325 (2012)
21. Yin, D., Hong, L., Xue, Z., Davison, B.D.: Temporal dynamics of user interests in tagging systems. In: Twenty-Fifth AAAI Conference on Artificial Intelligence (2011)
22. Yin, D., Hong, L., Davison, B.D.: Exploiting session-like behaviors in tag prediction. In: Proceedings of WWW'2011, pp. 167–168. ACM (2011)
23. Brainerd, C., Reyna, V.: Recollective and nonrecollective recall. J. Mem. Lang. 63, 425–445 (2010)
24. Kintsch, W., Mangalath, P.: The construction of meaning. Top. Cogn. Sci. 3, 346–370 (2011)
25. Krestel, R., Fankhauser, P.: Tag recommendation using probabilistic topic models. In: ECML PKDD Discovery Challenge 2009 (DC09), p. 131 (2009)
26. Lorince, J., Todd, P.M.: Can simple social copying heuristics explain tag popularity in a collaborative tagging system? In: Proceedings of WebSci '13, pp. 215–224. ACM, New York (2013)
27. Floeck, F., Putzke, J., Steinfels, S., Fischbach, K., Schoder, D.: Imitation and quality of tags in social bookmarking systems-collective intelligence leading to folksonomies. In: Bastiaens, T.J., Baumöl, U., Krämer, B.J. (eds.) On Collective Intelligence. AISC, vol. 76, pp. 75–91. Springer, Heidelberg (2010)
28. Seitlinger, P., Ley, T.: Implicit imitation in social tagging: familiarity and semantic reconstruction. In: Proceedings of CHI '12, pp. 1631–1640. ACM, New York (2012)
29. Helic, D., Körner, C., Granitzer, M., Strohmaier, M., Trattner, C.: Navigational efficiency of broad vs. narrow folksonomies. In: Proceedings of HT '12, pp. 63–72. ACM, New York (2012)
30. Gemmell, J., Schimoler, T., Ramezani, M., Christiansen, L., Mobasher, B.: Improving folkrank with item-based collaborative filtering. In: Recommender Systems & the Social Web (2009)
31. Doerfel, S., Jäschke, R.: An analysis of tag-recommender evaluation procedures. In: Proceedings of RecSys '13, pp. 343–346. ACM, New York (2013)
32. Campos, P.G., Díez, F., Cantador, I.: Time-aware recommender systems: a comprehensive survey and analysis of existing evaluation protocols. User Model. User-Adap. Inter. 24(1–2), 67–119 (2013)
33. Van Rijsbergen, C.J.: Foundation of evaluation. J. Doc. 30, 365–373 (1974)
34. Balby Marinho, L., Hotho, A., Jschke, R., Nanopoulos, A., Rendle, S., Schmidt-Thieme, L., Stumme, G., Symeonidis, P.: Recommender Systems for Social Tagging Systems. Springer Briefs in Electrical and Computer Engineering. Springer, Heidelberg (2012)
35. Kowald, D., Lacic, E., Trattner, C.: Tagrec: towards a standardized tag recommender benchmarking framework. In: Proceedings of HT'14. ACM, New York (2014)

36. Parra-Santander, D., Brusilovsky, P.: Improving collaborative filtering in social tagging systems for the recommendation of scientific articles. In: Proceedings of WI-IAT 2010, vol. 1, pp. 136–142. IEEE (2010)
37. Fu, W.T., Dong, W.: Collaborative indexing and knowledge exploration: a social learning model. IEEE Intell. Syst. **27**, 39–46 (2012)

Utilizing Online Social Network and Location-Based Data to Recommend Products and Categories in Online Marketplaces

Emanuel Lacic[1]([✉]), Dominik Kowald[1], Lukas Eberhard[2], Christoph Trattner[3], Denis Parra[4], and Leandro Balby Marinho[5]

[1] Know-Center, Graz University of Technology, Graz, Austria
{elacic,dkowald}@know-center.at
[2] IICM, Graz University of Technology, Graz, Austria
lukas.eberhard@tugraz.at
[3] Norwegian University of Science and Technology, Trondheim, Norway
chritrat@idi.ntnu.no
[4] Pontificia Universidad Catlica de Chile, Santiago, Chile
dparra@ing.puc.cl
[5] UFCG, Campina Grande, Brazil
lbmarinho@dsc.ufcg.edu.br

Abstract. Recent research has unveiled the importance of online social networks for improving the quality of recommender systems and encouraged the research community to investigate better ways of exploiting the social information for recommendations. To contribute to this sparse field of research, in this paper we exploit users' interactions along three data sources (marketplace, social network and location-based) to assess their performance in a barely studied domain: recommending products and domains of interests (i.e., product categories) to people in an online marketplace environment. To that end we defined sets of content- and network-based user similarity features for each data source and studied them isolated using an user-based Collaborative Filtering (CF) approach and in combination via a hybrid recommender algorithm, to assess which one provides the best recommendation performance. Interestingly, in our experiments conducted on a rich dataset collected from SecondLife, a popular online virtual world, we found that recommenders relying on user similarity features obtained from the social network data clearly yielded the best results in terms of accuracy in case of predicting products, whereas the features obtained from the marketplace and location-based data sources also obtained very good results in case of predicting categories. This finding indicates that all three types of data sources are important and should be taken into account depending on the level of specialization of the recommendation task.

Keywords: Recommender systems · Online marketplaces · SNA · Social data · Location-based data · SecondLife · Collaborative filtering · Item recommendations · Product recommendations · Category prediction

© Springer International Publishing Switzerland 2015
M. Atzmueller et al. (Eds.): MUSE/MSM 2013, LNAI 8940, pp. 96–115, 2015.
DOI: 10.1007/978-3-319-14723-9_6

1   Introduction

Research on recommender systems has gained tremendous popularity in recent years. Especially since the hype of the social Web and the rise of social media and networking platforms such as Twitter or Facebook, recommender systems are acknowledged as an essential feature helping users to, for instance, discover new connections between people or resources. Especially in the field of e-commerce sites, i.e., online marketplaces, current research is dealing with the improvement of the prediction task in order to recommend products that are more likely to match peoples preferences.

Typically, these online systems calculate personalized recommendations using only one data source, namely marketplace data (e.g., implicit user feedback such as previously viewed or purchased products - see also e.g., [1–3]). Although this approach has been well established and performs reasonably well, nowadays, online marketplaces often also have the opportunity to leverage additional information about the users coming from social and location-based data sources (e.g., via Facebook-connect). Even though previous research has shown that this kind of data can be useful in the wide field of recommender systems (see Sect. 2), it remains an open problem how to fully exploit these additional data sources (social network and location-based data) to improve the recommendation task in online marketplaces.

Moreover, it is often not the most important thing in online marketplaces to predict the exactly right products to the users but to suggest domains of interests (i.e., product categories) the users could like and could use for further browsing (e.g., [1,4]). Thus, it is not only important to investigate to which extent social and location-based data sources can be used to improve product recommendations but also to which extent this data can also be used for the recommendation of product categories.

To contribute to this sparse field of research, in this paper we present a first take on this problem in form of a research project that aims at understanding how different sources of interaction data can help in recommending products and categories to people in an online marketplace. In this respect, we are particularly interested in studying the efficiency of different user similarity features derived from various dimensions, not only from the marketplace but also from online social networks and location-based data to recommend products and categories to people via a user-based Collaborative Filtering (CF) approach (we have chosen a user-based CF approach since user-based CF is not only a well-established recommender algorithm but also allows us to incorporate various user-based similarity features coming from different data sources, which has been shown to play an important role in making more accurate predictions [4,5]). Specifically, we raise the following two research questions:

– *RQ1*: To which extent can user similarity features derived from marketplace, social network and location-based data sources be utilized for the recommendation of products and categories in online marketplaces?

- *RQ2*: Can the different marketplace, social network and location-based user similarity features and data sources be combined in order to create a hybrid recommender that provides more robust recommendations in terms of prediction accuracy, diversity and user coverage?

In order to address these research questions, we examined content-based and network-based user similarity feature sets for user-based CF over three data sources (marketplace, social and location-based data) as well as their combinations using a hybrid recommender algorithm and assessed the results via a more comprehensive set of recommender evaluation metrics than previous works. The study was conducted using a large-scale dataset crawled from the virtual world of SecondLife. In this way, we could study the utility of each user similarity feature separately as well as combine them in the form of hybrid approaches to show which combinations, per data source and globally, provide the best recommendations in terms of recommendation accuracy, diversity and user coverage. Summing this up, the contributions of this work are the following:

- Contrary to previous work in this area [1,2], we not only employ one source of data (marketplace) for the problem of predicting product purchases but show how data coming from three different sources (marketplace, social and location-based data) can be exploited in this context.
- In contrast to related work in the field, we provide also an extensive evaluation of various content-based and network-based user similarity features via user-based Collaborative Filtering as well as their combinations via a hybrid recommender approach.
- Finally, we also provide evidence to what extent top-level and sub-level purchase categories can be predicted which is in contrast to previous work (e.g., [1]) where the authors only focused on the problem recommending top-level categories to the users.

To the best of our knowledge, this is the first study that offers such a comprehensive user similarity feature selection and evaluation for product and category recommendation in online marketplaces.

Overall, the paper is structured as follows: we begin by discussing related work in Sect. 2. Then we present the datasets (Sect. 3) and the feature description (Sect. 4) used in our extensive evaluation. After that, we present our experimental setup in Sect. 5 and show the results of our experiments in Sect. 6. Finally, on Sect. 7 we conclude the paper and discuss the outlook.

2   Related Work

Most of the literature that leverages social data for recommendations is focused on recommending users, (e.g., [2,6]), tags (e.g., [7]) or points-of-interest (e.g., [4]), although some works have exploited social information for item or product recommendation, being the most important ones model-based. Jamali et al. [5]

introduced SocialMF, a matrix factorization model that incorporates social relations into a rating prediction task, decreasing RMSE with respect to previous work. Similarly, Ma et al. [4] incorporated social information in two models of matrix factorization with social regularization, with improvements in both MAE and RMSE for rating prediction. Among their evaluations, they concluded that choosing the right similarity feature between users plays an important role in making a more accurate prediction.

On a more general approach, Karatzoglou et al. [8] use implicit feedback and social graph data to recommend places and items, evaluating with a ranking task and reporting significant improvements over past related methods. Compared to these state-of-the-art approaches, our focus on this paper is at providing a richer analysis of feature selection (similarity features) with a more comprehensive evaluation than previous works, and in a rarely investigates domain: product recommendation in a social online marketplace. For instance, in Guo et al. [2] or Trattner et al. [3] the authors leveraged social interactions between sellers and buyers in order to predict sellers to customers. Other relevant work in this context is the study of Zhang and Pennacchiotti [1] who showed how top-level categories can be better predicted in a cold-start setting on eBay by exploiting the user's "likes" from Facebook.

3 Datasets

In our study we rely on three datasets[1] obtained from the virtual world Second-Life (SL). The main reason for choosing SL over real world sources are manifold but mainly due to the fact that currently there are no other datasets available that comprise marketplace, social and location data of users at the same time. For our study we focused on users who are contained in all three sources of data, which are 7,029 users in total. To collect the data (see Table 1) we crawled the SL platform as described in our previous work [9,10].

3.1 Marketplace Dataset

Similar to eBay, SecondLife provides an online trading platform where users can trade virtual goods with each other. Every seller in the SL marketplace[2] owns her own seller's store and publicly offers all of the store's products classified into different categories (a hierarchy with up to a maximum of four different categories per product). Furthermore, sellers can apply meta-data such as price, title or description to their products. Customers in turn, are able to provide reviews to products.

We extracted 29,802 complete store profiles, with corresponding 39,055 trading interactions, and 2,162,466 products, out of which 30,185 were purchased.

---

[1] **Note:** The datasets could be obtained by contacting the fourth author of this work.
[2] https://marketplace.secondlife.com/.

Table 1. Basic statistics of the SL datasets used in our study.

Marketplace dataset (Market)	
Number of products	$30,185$
Number of products with categories	$24,276$
Number of purchases	$39,055$
Number of purchases with categories	$31,164$
Mean number of purchases per user	5.56
Mean number of purchases per products	1.29
Mean number of categories per product	2.86
Number of top-level categories	23
Number of low-level categories	532
Mean number of top-level categories purchase	$1,354.96$
Mean number of low-level categories purchase	58.58
Number of sellers	$8,149$
Mean number of purchases per seller	3.70
Online social network dataset (Social)	
Number of interactions	$490,236$
Mean number of interactions	69.75
Number of groups	$39,180$
Mean number of groups per user	$9,419$
Number of interests	5.57
Mean number of interests per user	1.34
Location-based dataset (Location)	
Number of different favorite locations	$10,538$
Mean number of favorite locations per user	5.77
Number of different shared locations	$5,736$
Mean number of shared locations per user	1.94
Number of different monitored locations	$1,887$
Mean number of monitored locations per user	6.52

From the purchased products, 24,276 are described using categories which are differentiated in top-level categories and low-level categories (i.e., the assigned product categories on the lowest possible level of the category hierarchy). An example of a product in the marketplace of SL is shown in the first image of Fig. 1.

3.2 Online Social Network Dataset

The online social network MySecondLife[3] is similar to Facebook with regard to postings: users can interact with each other by sharing text messages and

---

[3] https://my.secondlife.com/.

(a) SecondLife store   (b) SecondLife social stream

Fig. 1. Examples for a store in the marketplace and a social stream of an user in the online social network of the virtual world SecondLife.

commenting or loving (= liking) these messages. From the extracted 7,029 complete user profiles, we gathered 39,180 different groups users belong to, 9,419 different interests users defined for themselves and 490,236 interactions between them. The second image of Fig. 1 shows an example of an user profile in the online social network of SL.

3.3   Location-Based Dataset

The world of Second Life is contained within regions, i.e., locations which are independent from each other. Overall, we extracted three different sources of location-based data in our experiments:

(a) Favored Locations: Every user of SL can specify up to 10 so-called "Picks" in their profile representing her favorite locations that other users can view in the user's MySecondLife profile. We found that the extracted users picked 40,558 locations from 10,538 unique locations;

(b) Shared Locations: Users in SL can also share information about their current in-world position through in-world pictures called "snapshots", which also include in-world GPS information (similar as Foursquare). Overall, we identified 13,637 snapshots in 5,736 unique locations;

(c) Monitored Locations: As in real life, users in SL can create events in the virtual world and publicly announce them in a public event calendar. We collected these events, with an accurate location and start time, and extracted 157,765 user-location-time triples, with 1,887 unique locations.

4   Feature Description

As shown in our previous work (e.g., [10]), similarities between users can be derived in two different ways: either we calculate similarities between users on the content (= meta-data) provided directly by the user profiles or on the network structure of the user profiles interacting with each other. In the following sections we present more details about these ideas.

4.1   Content-Based User Similarity Features

We define our set of content-based user similarity features based on different types of entities or meta-data information that are directly associated with the user profiles in our data sources. In the case of the marketplace dataset these entities are purchased products, product categories and sellers of the products, in the case of the social network these are groups and interests the users have assigned, and in the case of the location-based data source these are favored locations, shared locations and monitored locations. Formally, we define the entities of a user u as $\Delta(u)$ in order to calculate the similarity between two users, u and v.

The first content-based user similarity feature we induce is based on the entities two users have in common. It is called *Common Entities* and is given by:

$$sim(u, v) = |\Delta(u) \cap \Delta(v)| \tag{1}$$

The second similarity feature, *Total Entities*, is defined as the union of two users' entities and is calculated by:

$$sim(u, v) = |\Delta(u) \cup \Delta(v)| \tag{2}$$

These two user similarity features are combined by *Jaccard's Coefficient for Entities* as the number of common entities divided by the total number of entities:

$$sim(u, v) = \frac{|\Delta(u) \cap \Delta(v)|}{|\Delta(u) \cup \Delta(v)|} \tag{3}$$

4.2   Network-Based User Similarity Features

In our experiments we consider all networks as an undirected graph $G\langle V, E \rangle$ with V representing the user profiles and $e = (u, v) \in E$ if user u performed an action on v (see also [10]). In the case of the social network, these actions are defined as social interactions, which are a combination of likes, comments and wallposts. In the case of the location-based dataset, actions between users are determined if they have met each other in the virtual world at the same time in the same location[4]. Furthermore, the weight of an edge $w_{action}(u, v)$ gives the frequency of

---

[4] **Note:** We derived the networks in our study from the location-based dataset only for the monitored locations, since the exact timestamps are not available for the favored nor the shared locations in the datasets.

a specific action between two users u and v. Finally, this network structure also let us determine the neighbors of users in order to calculate similarities based on this information. We define the set of neighbors of a node $v \in G$ as $\Gamma(v) = \{u \mid (u, v) \in E\}$.

The first network-based user similarity feature we introduce, uses the number of *Directed Interactions* between two users and is given by:

$$sim(u, v) = w_{action}(u, v) \tag{4}$$

In contrast to *Directed Interactions*, the following user similarity features are based on the neighborhood of two users: The first neighborhood similarity feature is called *Common Neighbors* and represents the number of neighbors two users have in common:

$$sim(u, v) = |\Gamma(u) \cap \Gamma(v)| \tag{5}$$

To also take into account the total number of neighbors of the users, we introduced *Jaccard's Coefficient for Common Neighbors*. It is defined as:

$$sim(u, v) = \frac{|\Gamma(u) \cap \Gamma(v)|}{|\Gamma(u) \cup \Gamma(v)|} \tag{6}$$

A refinement of this feature was proposed as *Adamic Adar* [11], which adds weights to the links since not all neighbors in a network have the same tie strength:

$$sim(u, v) = \sum_{z \in \Gamma(u) \cap \Gamma(v)} \frac{1}{log(|\Gamma(z)|)} \tag{7}$$

Another related similarity feature introduced by Cranshaw et al. [12], called *Neighborhood Overlap*, measures the structural overlap of two users. Formally, this is written as:

$$sim(u, v) = \frac{|\Gamma(u) \cap \Gamma(v)|}{|\Gamma(u)| + |\Gamma(v)|} \tag{8}$$

The *Preferential Attachment Score*, first mentioned by Barabasi et al. [13], is another network-based similarity feature with the goal to prefer active users in the network. This score is the product of the number of neighbors of each user and is calculated by:

$$sim(u, v) = |\Gamma(u)| \cdot |\Gamma(v)| \tag{9}$$

5   Experimental Setup

In this section we provide a detailed description of our experimental setup. First, we describe the recommender approaches we have chosen in order to evaluate our three data sources (marketplace, social network and location-based data) as well as our derived user similarity features for the task of recommending products and categories. Afterwards, we describe the evaluation methodology and the metrics used in our study.

5.1   Recommender Approaches

In this subsection we describe the recommender approaches we have used to tackle our research questions described in the introductory section of this paper. All mentioned approaches have been implemented into our scalable big data social recommender framework *SocRec* [9,14], an open-source framework which can be obtained from our Github repository[5].

Baseline. As baseline for our study, we used a simple MostPopular approach recommending the most popular products or categories in terms of purchase frequency to the users.

Recommending Products. The main approach we adopt to evaluate our data sources and user similarity features for the task of recommending products is a *User-based Collaborative Filtering (CF)* approach. The basic idea of this approach is that users who are more similar to each other, e.g., have similar taste, will likely agree or rate on other resources in a similar manner [15]. Out of the different CF approaches, we used the non-probabilistic user-based nearest neighbor algorithm where we first find the k-nearest similar users and afterwards recommend the resources of those user as a ranked list of top-N products to the target user that are new to her (i.e., she has not purchased those products in the past). As outlined before, we have chosen this approach since user-based CF is not only a well-established recommender algorithm but also allows us to incorporate various user-based similarity features coming from different data sources, which has been shown to play an important role in making more accurate predictions [4,5].

The similarity values of the user pairs $sim(u, v)$ are calculated based on the user similarity features proposed in Sect. 4 (i.e., constructing the neighborhood). Based on these similarity values, each item i of the k most similar users for the target user u is ranked using the following formula [15]:

$$pred(u, i) = \sum_{v \in neighbors(u)} sim(u, v) \qquad (10)$$

Recommending Categories. For the task of recommending categories and in contrast to previous work (e.g., [1]), we are not only focusing here on the prediction of top-level categories but also on the prediction of low-level categories. The prediction of categories was implemented as an extension of product predictions. Thus, for each product in the list of recommended products (i.e., the products obtained from the k-nearest neighbors of user u based on a user similarity feature), we extracted the assigned category on the highest level in the case of predicting top-level categories and the assigned category on the lowest level in the case of predicting low-level categories. Afterwards, we assigned a score to

---

[5] https://github.com/learning-layers/SocRec.

each extracted category e_i in the set of all extracted categories E_u for the target user u based on a similar method as proposed in [1]:

$$pred(u, e_i) = \frac{purc(E_u, e_i)}{\sum\limits_{e \in E_u} purc(E_u, e)} \qquad (11)$$

where $purc(E_u, e_i)$ gives the number of times the category e_i occurs in the set of all extracted categories E_u for user u.

Combining User Similarity Features and Data Sources. To further explore how to combine our data sources and features for recommendation, we investigated different hybridization methods (see also [16]). The hybrid approach chosen in the end is known as *Weighted Sum*. The score of each recommended item in the *Weighted Sum* algorithm is calculated as the weighted sum of the scores for all recommender approaches. It is given by:

$$W_{rec_i} = \sum\limits_{s_j \in S} (W_{rec_i, s_j} \cdot W_{s_j}) \qquad (12)$$

where the combined weighting of the recommended item i, W_{rec_i}, is given by the sum of all single weightings for each recommended item in an approach W_{rec_i, s_j} multiplied by the weightings of the recommender approaches W_{s_j}. We weighted each recommender approach W_{s_j} based on the nDCG@10 value obtained from the individual approaches (calculated based on an additional evaluation set where we withheld 20 purchased products - see Sect. 5.2).

We also experimented with other hybrid approaches, known as *Cross-source* and *Mixed Hybrid* [16]. However, these approaches have not yielded better results than the *Weighted Sum* algorithm.

5.2   Evaluation Method and Metrics

To evaluate the performance of each approach in a recommender setting, we performed a number of off-line experiments. Therefore, we split the SL dataset in two different sets (training and test set) using a method similar to the one described in [9], i.e., for each user we withheld 10 purchased products from the complete dataset and added them to the test set to be predicted. Since we did not use a p-core pruning technique to prevent a biased evaluation, there are also users with less than 10 relevant products. We did not include these users into our evaluation since they did not have enough purchase data available that could be used to produce reasonable recommendations based on the marketplace data, although this data is worthwhile for our user-based CF approach to find suitable neighbors. Thus, we used a post-filtering method, where all the recommendations were still calculated on the whole datasets but accuracy estimates were calculated only based on these filtered user profiles (= 959 users).

To finally quantify the performance of each of our recommender approaches, we used a diverse set of well-established metrics in recommender systems [17,18]. These metrics are as follows:

Recall (R@k) is calculated as the number of correctly recommended products divided by the number of relevant products, where r_u^k denotes the top k recommended products and R_u the list of relevant products of a user u in the set of all users U. Recall is given by [19]:

$$R@k = \frac{1}{|U|} \sum_{u \in U} \left(\frac{|r_u^k \cap R_u|}{|R_u|} \right) \tag{13}$$

Precision (P@k) is calculated as the number of correctly recommended products divided by the number of recommended products k. Precision is defined as [19]:

$$P@k = \frac{1}{|U|} \sum_{u \in U} \left(\frac{|r_u^k \cap R_u|}{k} \right) \tag{14}$$

Normalized Discounted Cumulative Gain (nDCG@k) is a ranking-dependent metric that not only measures how many products can be correctly predicted but also takes the position of the products in the recommended list with length k into account. The nDCG metric is based on the *Discounted Cummulative Gain (DCG@k)* which is given by [20]:

$$DCG@k = \sum_{k=1}^{|r_u^k|} \left(\frac{2^{B(k)} - 1}{log_2(1 + k)} \right) \tag{15}$$

where $B(k)$ is a function that returns 1 if the recommended product at position i in the recommended list is relevant. nDCG@k is calculated as DCG@k divided by the ideal DCG value iDCG@k which is the highest possible DCG value that can be achieved if all the relevant products would be recommended in the correct order. Taken together, it is given by the following formula [20]:

$$nDCG@k = \frac{1}{|U|} \sum_{u \in U} \left(\frac{DCG@k}{iDCG@k} \right) \tag{16}$$

Diversity (D@k), as defined in [17], can be calculated as the average dissimilarity of all pairs of resources in the list of recommended products r_u^k. Given a distance function $d(i, j)$ that is the distance, or the dissimilarity between two products i and j, D is given as the average dissimilarity of all pairs of products in the list of recommended products [17]:

$$D@k = \frac{1}{|U|} \sum_{u \in U} \left(\frac{1}{k \cdot (k-1)} \sum_{i \in R} \sum_{j \in r_u^k, j \neq i} d(i, j) \right) \tag{17}$$

User Coverage (UC) is defined as the number of users for whom at least one product recommendation could have been calculated ($|U_r|$) divided by total number of users $|U|$ [21]:

$$UC = \frac{|U_r|}{|U|} \tag{18}$$

All mentioned performance metrics are calculated and reported based on the top-10 recommended products.

6   Results

In this section we highlight the results of our experiments for predicting products, low-level and top-level categories in terms of algorithmic performance in order to tackle our two research questions presented in Sect. 1. Our evaluation has been conducted in two steps: in the first step we compared the different recommender approaches with the corresponding user similarity features isolated ($RQ1$), see Table 2) and in the second step we combined these approaches in the form of hybrid recommendations ($RQ2$, see Table 3). All results are presented by recommender accuracy, given by nDCG@10, P@10 (Precision) and R@10 (Recall), D@10 (Diversity) and UC (User Coverage).

6.1   Recommendations Based on Single User Similarity Features

The results for the recommendation of products, low-level categories and top-level categories, using content-based and network-based user similarity features derived from our three data sources (marketplace, social and location-based data), are shown in Table 2 in order to address our first research question ($RQ1$). The results also include the *Most Popular (MP)* approach as a baseline. Additionally, the performance of the different data sources is also shown in Fig. 2 in form of Recall/Precision plots.

Recommending Products. Regarding the task of predicting product purchases (first column in Table 2), the best results in terms of recommender accuracy are reached by the network-based features based on interactions (e.g., loves, comments, wallposts) between the users in the social network. Surprisingly these approaches clearly outperform the user-based CF approaches relying on marketplace data, which implies that social interactions of the users are a better predictor to recommend products to people than marketplace data.

Another interesting finding is that the neighborhood-based features (*Common Neighbors, Jaccard, Neighborhood Overlap* and *Adamic/Adar*) also seem to be better indicators to determine the similarity between users than the direct interactions between these pairs. Only the *Preferential Attachment Score* based recommender approach does not perform well in this context, although it still performs better than the features derived from the marketplace data source. This is to some extent expected and reveals that the individual's taste is more driven by the user's peers rather than by the popular users in the SecondLife social network.

In terms of the marketplace and location-based user similarity features, the results reveal that they do not provide high estimates of accuracy. This is

Table 2. Results for the user-based CF approaches based on various user similarity features showing their performance for the tasks of predicting products, low-level categories and top-level categories, respectively (*RQ1*). *Note:* Bold numbers indicate the highest accuracy values per feature set and "*" indicate the overall highest accuracy estimates.

	User Sim. Feature	Products			low-level categories			top-level categories				
		nDCG@10	P@10	R@10	nDCG@10	P@10	R@10	nDCG@10	P@10	R@10	D@10	UC
	Most Popular	.0048	.0037	.0047	.0185	.0207	.0157	.2380	.2730	.2221	.6392	100.0%
Market / Content	Common Purchases	.0097	.0073	.0094	.0724	.0641	.0757	.4557	.3884	.4636	.5892	90.51%
	Common Sellers	.0146	.0102	.0142	.1119	.1005	.1132	.5251	.4610	.5183	.6372	99.06%
	Jaccard Sellers	**.0158**	**.0114**	**.0154**	.1092	.1029	.1047	.5061	.4940	.4927	.6054	99.06%
	Total Sellers	.0065	.0052	.0073	.0929	.0743	.0977	.5079	.4094	.5113	.6566	99.06%
	Common Categories	.0050	.0039	.0054	.1090	.1051	.1041	.5073	.5366	.4674	.6123	99.48%
	Jaccard Categories	.0058	.0039	.0049	**.1361**	**.1301**	**.1288**	**.5456***	**.5701***	**.5200***	.6364	99.48%
	Total Categories	.0007	.0006	.0009	.0225	.0236	.0280	.3317	.3353	.4215	.6575	99.48%
Social / Content	Common Groups	.0022	.0010	.0014	.0402	.0320	.0425	.3233	**.2567**	.3439	.4307	64.13%
	Jaccard Groups	**.0027**	**.0016**	**.0021**	**.0433**	**.0339**	**.0459**	**.3272**	.2557	**.3464**	.4332	64.13%
	Total Groups	.0006	.0005	.0007	.0324	.0272	.0361	.3214	.2323	.3563	.4466	64.13%
	Common Interests	.0005	.0002	.0002	.0235	.0201	.0267	.2285	.1810	.2474	.3185	46.51%
	Jaccard Interests	.0003	.0002	.0003	.0219	.0201	.0227	.2424	.1857	.2551	.3161	46.51%
	Total Interests	.0004	.0002	.0003	.0285	.0235	.0337	**.2639**	.1793	**.2765**	.3319	46.51%
Social / Network	Directed Interactions	.0345	.0389	.0471	.0582	.0528	.0630	.1743	.1756	.1769	.2169	38.79%
	Common Neighbors	.1107	.1021	.1104	.1216	.1212	.1243	.2633	.2887	.2584	.3300	62.88%
	Jaccard Common Neighbors	.1381	.1143	.1378	.1523	.1386	.1618	.3726	.3419	.3814	.4455	71.53%
	Neighborhood Overlap	**.1434***	**.1222***	**.1471***	**.1620***	**.1486***	**.1719***	**.3855**	**.3471**	**.3965**	.4514	71.53%
	Adamic/Adar	.1013	.0941	.1067	.1241	.1187	.1272	.3028	.3153	.3042	.3762	69.34%
	Pref. Attach. Score	.0317	.0331	.0380	.0630	.0569	.0650	.3202	.2838	.3332	.4420	70.70%
Location / Content	Common Favored Locations	.0019	.0010	.0015	.0427	.0393	.0481	.4674	.3773	.4946	.6437	96.35%
	Jaccard Favored Locations	.0028	**.0017**	.0022	.0473	.0416	.0531	.4636	.3777	.4919	.6490	96.35%
	Total Favored Locations	**.0031**	.0016	**.0023**	.0459	.0400	.0513	.4794	.3802	.5061	.6635	96.35%
	Common Shared Locations	.0003	.0003	.0004	.0130	.0103	.0144	.1449	.1180	.1599	.2067	30.45%
	Jaccard Shared Locations	.0005	.0003	.0004	.0134	.0115	.0145	.1420	.1208	.1522	.2042	30.45%
	Total Shared Locations	.0000	.0000	.0000	.0092	.0074	.0106	.1340	.1076	.1520	.2031	30.45%
	Common Monitored Locations	.0016	.0010	.0014	.0408	.0345	.0449	.4825	**.3804**	.5139	.6734	98.23%
	Jaccard Monitored Locations	.0017	.0008	.0012	**.0473**	**.0403**	**.0546**	**.4987**	.3795	**.5387**	.6760	98.23%
	Total Monitored Locations	.0011	.0006	.0009	.0366	.0331	.0442	.4770	.3703	.5354	.6757	98.23%
Location / Network	Common Neighbors	.0015	.0007	.0010	.0298	.0271	.0345	.3377	.2623	.3632	.4609	67.05%
	Jaccard Common Neighbors	**.0016**	**.0008**	**.0011**	**.0322**	.0261	**.0367**	.3306	.2623	.3545	.4579	67.05%
	Neighborhood Overlap	.0014	.0007	.0010	.0295	.0267	.0339	.3359	.2614	.3608	.4615	67.05%
	Adamic/Adar	.0015	.0006	.0009	.0320	**.0272***	.0353	.3332	.2598	.3530	.4595	67.05%
	Pref. Attach. Score	.0003	.0002	.0002	.0270	.0213	.0335	**.3634**	**.2825**	.3458	.4583	70.59%

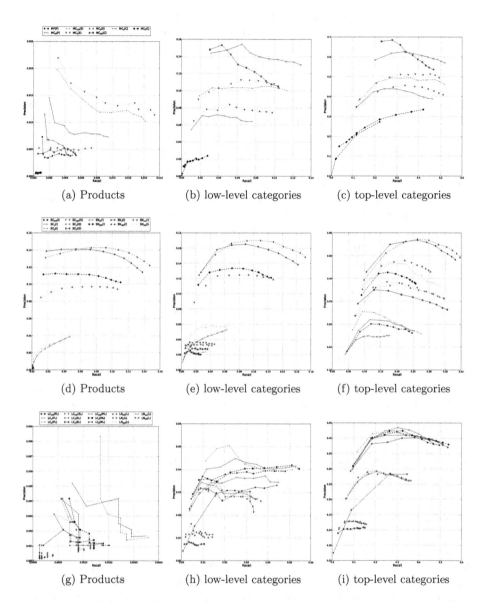

(a) Products

(b) low-level categories

(c) top-level categories

(d) Products

(e) low-level categories

(f) top-level categories

(g) Products

(h) low-level categories

(i) top-level categories

Fig. 2. Recall/Precision plots for the single user similarity features derived from the marketplace (a, b, c), social (d, e, f) and location-based (g, h, i) data sources, showing the performance of each feature for $k = 1$–10 recommended items, low-level categories or top-level categories, respectively (*RQ1*). *Note:* Each feature name in the legends is derived in the following way: the first two letters describe the data source, the subscript denotes the user similarity feature and the value in brackets defines the used data field (e.g., $SN_{NO}(I)$ stands for the Social Network data source, the Neighborhood Overlap similarity feature and Interactions data field).

interesting since our previous work [10, 22] showed that these features perform extremely well in predicting tie strength or social interactions between users. However, the features derived from the marketplace and the location-based data sources provide the best results with respect to Diversity (D) and User Coverage (UC).

Recommending Categories. Regarding the tasks of predicting low-level and top-level categories, the second and third column of Table 2 report the accuracy estimates for the different user similarity features based on the extracted categories. As expected, all user similarity features end up with a much higher accuracy than for predicting products, especially in the case of top-level categories, because of the lower level of specialization of these recommendation tasks. In the case of the low-level category predictions, the approaches based on social interaction features still perform better than the approaches based on features of the marketplace or location-based data sources. Interestingly, the content-based user similarity features derived from the social network as well as the location-based features, which performed the worst at product predictions, perform much better for low-level categories, now also outperforming the *MP* baseline.

In the case of the top-level category recommendations, it can be seen that the user similarity features of all three data sources provide quite similar results in terms of recommender accuracy. The approach based on the Jaccard's coefficient for categories performs best in terms of nDCG@10, P@10 and R@10. This result is very interesting since this feature is based on the marketplace data source that provided quite bad results in the case of product predictions. Summed up, we see that user similarity features derived from all data sources are very useful indicators for recommendations, although they depend on the level of specialization of the recommendation task (*RQ1*).

6.2 Recommendations Based on Combined Data Sources

The findings of the last subsection suggests that a combination of features from all three data sources (marketplace, social network and location-based data) should provide more robust recommendations in case of both tasks, predicting products and categories (*RQ2*). Thus, Table 3 shows the results of the hybrid approaches based on theses data sources in order to tackle our second research question. As before, the first column indicates the results for the product prediction, the second column for the low-level category prediction and the third column for the top-level category prediction. Additionally, Fig. 3 shows the performance of the hybrid approaches based on the three data sources for $k = 1$–10 recommended products, low-level categories or top-level categories, respectively, in form of Recall/Precision plots.

Recommending Products. Regarding the product prediction task, we see again that the recommender approaches based on the social network data source clearly outperform the ones based on the marketplace and location-based data sources as well as the *MP* baseline. Furthermore, when combining all three data sources, not only the overall recommendation accuracy is increased with respect

Table 3. Results for the hybrid approaches based on our three data sources for the tasks of predicting products, low-level categories and top-level categories. The results show that all three data sources (marketplace, social network and location-based data) are important indicators for calculating recommendations, since a hybrid combination of all data sources provided the best results in case of predicting products, low-level categories and top-level categories (RQ2). *Note:* Bold numbers indicate the highest accuracy values across the data sources and "*" indicate the overall highest accuracy estimates.

	Sets		Products			low-level categories			top-level Categories				
			nDCG@10	P@10	R@10	nDCG@10	P@10	R@10	nDCG@10	P@10	R@10	D@10	UC
Weighted Sum	Most Popular		.0082	.0021	.0122	.0185	.0207	.0157	.2380	.2730	.2221	.5945	100.00%
	Market	Content	.0151	.0103	.0142	.1051	.0886	.1046	.5232	**.4476**	.5210	.6421	99.79%
	Social	Content	.0030	.0019	.0025	.0702	.0432	.0571	.5555	.3274	.4523	.5603	81.65%
		Network	.1416	**.1180**	.1450	**.2454***	.1416	.1740	.5708	.3385	.4047	.4591	71.53%
		Combined	.1418	.1165	**.1454**	.2104	**.1515**	**.1942**	**.5901**	.4131	.5446	.6245	92.70%
	Location	Content	.0036	.0021	.0030	.0556	.0406	.0556	.5535	.3870	**.5535**	.6923	100.00%
		Network	.0015	.0007	.0010	.0535	.0284	.0378	.5359	.2672	.3783	.4864	70.59%
		Combined	.0036	.0022	.0031	.0540	.0406	.0540	.5497	.3832	.5497	.6914	100.00%
	Combined		**.1460***	**.1187***	**.1493***	.2163	.1708	.2159	**.5978***	.4656	**.5965***	.6642	100.00%
	Combined Top 3		.1459	.1186	.1475	.2161	**.1801***	**.2161***	.5829	**.4898***	.5829	.6530	100.00%

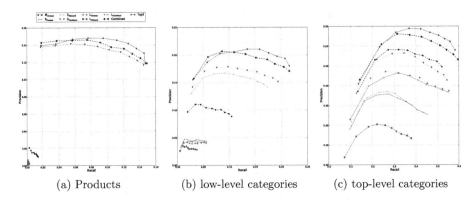

(a) Products   (b) low-level categories   (c) top-level categories

Fig. 3. Recall/Precision plots for the hybrid approaches showing the performance of each data source for $k = 1$–10 recommended products, low-level categories or top-level categories, respectively (*RQ2*).

to nDCG@10, P@10 and R@10, but also the User Coverage (UC) is increased to the maximum of 100 %.

This means that the hybrid approach combines the strengths of the user similarity features of all three data sources in order to be capable of providing accurate recommendations for all users in the datasets. Another hybrid approach shown in Table 3 combines only the best user similarity features from each data source (referred to as *Top 3*) and reaches higher accuracy estimates, but lower Diversity (D).

Recommending Categories. In contrast to the results of the product predictions, that showed that the recommender based on the social network data source clearly outperforms the recommenders based on the marketplace and location-based data sources, the results of the category predictions (second and third column of Table 3) do not show that big differences between the three data sources. With respect to the low-level category predictions, we again observe that the recommender based on the social network data source still provides the highest accuracy estimates.

Interestingly, in this case also the recommenders based on the other two data sources provide reasonable results, which has not been the case of predicting products, where the location-based recommender even was outperformed by the *MP* baseline. Based on these results we would assume that marketplace and location-based data sources are suitable of providing accurate predictions in more general recommendation tasks. The results for the top-level category predictions prove this assumption since in this case the recommenders based on marketplace and location-based data sources even provide better results in terms of recommender accuracy, Diversity and User Coverage than the one based on social network data in case of the content-based features. As before, the combination of all three data sources provide again the best results. Summed up, this results prove our assumption derived from *RQ1*, that all three data sources are important for calculating recommendations, since a combination of all data sources provided the best results in case of predicting products, low-level categories and top-level categories (*RQ2*).

7   Conclusions and Future Work

In this work we presented first results of a recently started project that tries to utilize various user similarity features derived from three data sources (marketplace, social network and location-based data) to recommend products and points of interests (i.e., low-level and top-level categories) to people in an online marketplace setting. This section concludes the paper with respect to our two research questions and gives an outlook into the future.

The first research question of this work (*RQ1*) dealt with the question as to which extent user similarity features derived from marketplace, social network and location-based data sources can be utilized for the recommendation of products and categories in online marketplaces. To tackle this question we implemented various user-based Collaborative Filtering (CF) approaches based on the user similarity features from the data sources and tested them isolated. As the results have shown, the user-based CF approaches that utilize features of online social network data to calculate the similarities between users performed best in case of predicting products, significantly outperforming the other approaches relying on both – marketplace and location-based user data. However, this behavior changed in the case of predicting low-level and top-level categories where the differences between the three data sources got substantially smaller. Surprisingly, with respect to the top-level category predictions, the marketplace and location-based features even reached the highest results in terms of accuracy, Diversity (D) and User Coverage (UC).

These results showed that user similarity features of all three data sources are important indicators for recommendations and suggests that combining them should result into more robust recommendations, especially in cases of multiple recommendation tasks on different levels of specialization (topics and categories). Thus, our second research question (*RQ2*) tried to tackle the question if the different marketplace, social network and location-based user similarity features and data sources can be combined in order to create a hybrid recommender that provides more robust recommendations in terms of prediction accuracy, diversity and user coverage. In order to address this question we implemented and evaluated hybrid recommenders that combined the features of the data sources. The results proved our assumption and showed that hybrid recommender that combined user similarity features of all three data sources provided the best results across all accuracy metrics (nDCG@10, P@10, R@10) and all settings (product, low-level category and top-level category recommendations). Moreover, this hybrid recommender also provided a User Coverage of 100 % and thus, is able to provide these accurate recommendation to all users in the datasets.

Although the results of this study are based on a dataset obtained from the virtual world SecondLife, we believe that it bears great potential to create a sequence of interesting studies that may have implications for the "real" world (see e.g., [23]). For instance, one of the potential interesting issues we are currently exploring is predicting products and categories to users in a cold-start setting (i.e., for users that only have purchases a few or even no products in the past) by a diversity of features. Other important work we plan is the use of state-of-the-art model-based approaches in order to assess whether the signals extracted from similarity features in the current analysis can be replicated (in the case of social data) or improved (in the case of location data) for different recommendation tasks.

We have also shown that using the interaction information between users improves not only the task of product recommendation, but also the recommendation of low-level and top-level categories. Thus, we are also interested in studying the extent to which recommendations can be improved by utilizing content-based similarity features derived from the users' social streams.

Acknowledgments. This work is supported by the Know-Center and the EU funded project Learning Layers (Grant Agreement 318209). Moreover, parts of this work were carried out during the tenure of an ERCIM "Alain Bensoussan" fellowship programme. The Learning Layers project is supported by the European Commission within the 7th Framework Program, under the DG Information society and Media (E3), unit of Cultural heritage and technology-enhanced learning. The Know-Center is funded within the Austrian COMET Program - Competence Centers for Excellent Technologies - under the auspices of the Austrian Ministry of Transport, Innovation and Technology, the Austrian Ministry of Economics and Labor and by the State of Styria. COMET is managed by the Austrian Research Promotion Agency (FFG).

References

1. Zhang, Y., Pennacchiotti, M.: Predicting purchase behaviors from social media. In: Proceedings of WWW '13, pp. 1521–1532 (2013)
2. Guo, S., Wang, M., Leskovec, J.: The role of social networks in online shopping: Information passing, price of trust, and consumer choice. In: Proceedings of EC '11, pp. 157–166. ACM (2011)
3. Trattner, C., Parra, D., Eberhard, L., Wen, X.: Who will trade with whom? Predicting buyer-seller interactions in online trading platforms through social networks. In: Proceedings of WWW '14, pp. 387–388. ACM (2014)
4. Ma, H., Zhou, D., Liu, C., Lyu, M.R., King, I.: Recommender systems with social regularization. In: Proceedings of WSDM '11, pp. 287–296. ACM (2011)
5. Jamali, M., Ester, M.: A matrix factorization technique with trust propagation for recommendation in social networks. In: Proceedings of RecSys '10, pp. 135–142. ACM, New York (2010)
6. Bischoff, K.: We love rock'n'roll: analyzing and predicting friendship links in Last.fm. In: Proceedings of WebSci '12, pp. 47–56. ACM (2012)
7. Feng, W., Wang, J.: Incorporating heterogeneous information for personalized tag recommendation in social tagging systems. In: Proceedings of KDD '12, pp. 1276–1284. ACM (2012)

8. Delporte, J., Karatzoglou, A., Matuszczyk, T., Canu, S.: Socially enabled preference learning from implicit feedback data. In: Blockeel, H., Kersting, K., Nijssen, S., Železný, F. (eds.) ECML PKDD 2013, Part II. LNCS, vol. 8189, pp. 145–160. Springer, Heidelberg (2013)
9. Lacic, E., Kowald, D., Parra, D., Kahr, M., Trattner, C.: Towards a scalable social recommender engine for online marketplaces: The case of apache solr. In: Proceedings of WWW '14, pp. 817–822. ACM (2014)
10. Steurer, M., Trattner, C.: Acquaintance or partner? Predicting partnership in online and location-based social networks. In: Proceedings of ASONAM'13. IEEE/ACM (2013)
11. Adamic, L., Adar, E.: Friends and neighbors on the web. Soci. Netw. **25**, 211–230 (2003)
12. Cranshaw, J., Toch, E., Hong, J., Kittur, A., Sadeh, N.: Bridging the gap between physical location and online social networks. In: Proceedings of the 12th ACM International Conference on Ubiquitous Computing, pp. 119–128. ACM (2010)
13. Barabási, A., Albert, R.: Emergence of scaling in random networks. Science **286**, 509–512 (1999)
14. Lacic, E., Kowald, D., Trattner, C.: Socrecm: A scalable social recommender engine for online marketplaces. In: Proceedings of HT '14, pp. 308–310 (2014)
15. Schafer, J.B., Frankowski, D., Herlocker, J., Sen, S.: Collaborative filtering recommender systems. In: The adaptive web. Springer (2007) 291–324.
16. Bostandjiev, S., O'Donovan, J., Höllerer, T.: Tasteweights: a visual interactive hybrid recommender system. In: Proceedings of RecSys '12, pp. 35–42. ACM (2012)
17. Smyth, B., McClave, P.: Similarity vs. Diversity. In: Aha, D.W., Watson, I. (eds.) ICCBR 2001. LNCS (LNAI), vol. 2080, pp. 347–361. Springer, Heidelberg (2001)
18. Herlocker, J.L., Konstan, J.A., Terveen, L.G., Riedl, J.T.: Evaluating collaborative filtering recommender systems. ACM Trans. Inf. Syst. (TOIS) **22**, 5–53 (2004)
19. Van Rijsbergen, C.J.: Foundation of evaluation. J. Doc. **30**, 365–373 (1974)
20. Parra, D., Sahebi, S.: Recommender systems: sources of knowledge and evaluation metrics. In: Velásquez, J.D., Palade, V., Jain, L.C. (eds.) Advanced Techniques in Web Intelligence-2. SCI, vol. 452, pp. 149–176. Springer, Heidelberg (2013)
21. Ge, M., Delgado-Battenfeld, C., Jannach, D.: Beyond accuracy: evaluating recommender systems by coverage and serendipity. In: Proceedings of the Fourth ACM Conference on Recommender Systems, pp. 257–260. ACM (2010)
22. Steurer, M., Trattner, C.: Predicting interactions in online social networks: an experiment in second life. In: Proceedings of the 4th International Workshop on Modeling Social Media, p. 5. ACM (2013)
23. Szell, M., Sinatra, R., Petri, G., Thurner, S., Latora, V.: Understanding mobility in a social petri dish. Scientific Reports 2 (2012)

Open Smartphone Data for Structured Mobility and Utilization Analysis in Ubiquitous Systems

Nico Piatkowski[1(✉)], Jochen Streicher[2], Olaf Spinczyk[2], and Katharina Morik[1]

[1] Department of Computer Science, LS8, TU Dortmund University,
44227 Dortmund, Germany
{nico.piatkowski,katharina.morik}@tu-dortmund.de
http://www-ai.cs.uni-dortmund.de

[2] Department of Computer Science, LS12, TU Dortmund University,
44227 Dortmund, Germany
{jochen.streicher,olaf.spinczyk}@tu-dortmund.de
http://ess.cs.uni-dortmund.de

Abstract. The development and evaluation of new data mining methods for ubiquitous environments and systems requires real data that were collected from real users. In this work, we present an open smartphone utilization and mobility data set that was generated with several devices and participants during a 4-month study. A particularity of this data set is the inclusion of low-level operating system data. Additionally to the description of the data, we also describe the process of collection and the privacy measures we applied. To demonstrate the utility of the data, we evaluate the quality of generative spatio-temporal models for "apps" and network cells, since these are required as a building block in general predictions of the resource consumption of ubiquitous systems.

1 Introduction

Today's mobile phones are able to produce a vast amount of valuable data. Produced by several physical and logical sensors, the data provides knowledge about the owner as well as his environment. Several studies have shown that smartphones can be used as an effective tool to gain insights into patterns of human behavior and interaction that were not available before. Notable examples are the data sets of the MIT Human Dynamics Lab like the Reality Mining data set [1] or the Lausanne Data Collection Campaign [2]. The latter was however only available for participants of the 2012 Nokia data challenge [3].

While smartphones are certainly an excellent *tool* for research, it is not less important to consider how data collection campaigns can help to improve these devices and the respective infrastructure. Previous research has shown that insights into utilization and mobility patterns of mobile devices are indeed of value for that purpose. This concerns the mobile network infrastructure [4] as well as the user experience with respect to the devices. A limiting factor to user

© Springer International Publishing Switzerland 2015
M. Atzmueller et al. (Eds.): MUSE/MSM 2013, LNAI 8940, pp. 116–130, 2015.
DOI: 10.1007/978-3-319-14723-9_7

experience is certainly the lifetime of a smartphone's battery. Much research has been conducted towards the use of user-specific mobility and utilization patterns to increase the energy-efficiency of mobile devices, like the reduction of GPS utilization via location prediction [5], replacing energy-demanding sensors with correlated utilization data for context inference [6], energy-aware cellular data scheduling [7], or accelerated file prefetching [8].

Research on these problems requires real data collected on real devices and from real users. While high-level data like location and phone call logs might be sufficient for some of the problems, others also need data concerning the device, not only its owner. *Device Analyzer*[1] is an Android application collecting data from thousands of devices all over the world in order to get insights into utilization patterns, with the explicit goal to provide crucial information for the improvement of future smartphones. To guarantee complete anonymity all privacy-critical identifiers (e.g., cell tower IDs or MAC-addresses) are hashed with individual salts. This makes the data set unavailable for the analysis of social interaction. Also, it is not clear when the data will be available.

Our proposed data set shares commonalities with all three mentioned examples, but features all of the following: (1) It contains operating system level data, (2) not all identifiers are hashed with individual salts, and (3) it is openly available at http://sfb876.tu-dortmund.de/mobidata.

The remainder of this paper is structured as follows: In Sect. 2 we will shortly describe how we collected the data and ensured the privacy of our participants, while Sect. 3 describes the resulting data set. In Sect. 4 we present three exemplary analyses performed on these data. Section 5 concludes the paper.

2   Collection Process

We started with 11 participants, who all were members of our collaborative research center. During our summer school in 2012, we additionally collected data from 11 attendees. For this purpose we used *MobiDAC*, our flexible infrastructure for data collection on Android-based smartphones. MobiDAC allows experimenters to use the participating devices like programmable sensor nodes. Experimenters write *sensing modules* that perform the actual data acquisition on the device. These modules may be dynamically and remotely started, stopped and uploaded to a new participants' devices. The permission to do so is controlled by the respective participants. When a module is running, it is collecting, possibly preprocessing and saving data locally on the device. Data is sent back to the experimenter when certain conditions are met, like an established Wi-Fi connection. Currently, a modified version of the Scripting Layer for Android (SL4A)[2] is used to execute the sensing modules.

---

[1] Device Analyzer website: http://deviceanalyzer.cl.cam.ac.uk/.

[2] SL4A can be found at: http://code.google.com/p/android-scripting.

2.1   Modus Operandi

We used both the Android-API as well as the Linux virtual file systems (VFS) "/proc" and "/sys" as data sources. When the device was awake, most of the data was collected high-frequently (temporal resolution of two seconds) or via callbacks from Android. To reduce the amount of data that had to be transmitted from the device, we only recorded changes to data values. Every 60 s, we took a snapshot of all data from the virtual file systems and started sensor sampling for two seconds with the highest possible frequencies. A bluetooth scan was started every five minutes. Every two hours, the periodically sampled data was recorded completely (not only changes). As opposed to Symbian-based phones, which were used for some of the data collection campaigns mentioned in the introduction, Android phones try to *suspend* whenever possible. This happens, when the screen is off and no application is keeping the device awake by means of a *wake lock*. During the suspended state, no data can be collected at all. Thus, we explicitly wake the device from its sleep every 60 s and perform a full acquisition of all data values with the respective intervals.

2.2   Privacy Preservation

Since this version of our data set is truly open and available to anyone, we were obliged to be especially careful in the process of ensuring privacy. We treated data in the following ways:

- Everything that uniquely identifies a participant is *globally consistently* replaced with a random value. This is also true for all identifiers from inter-action with other entities (e.g., MAC-addresses and SSIDs) as well as for the names of installed and running application packages and processes.
- Mobile network cell information was replaced by *locally consistent* random values for each participant. This means that the mapping of cell identification (CID) and location area code (LAC) is different for every participant.

3   Data

We collected data from various hardware and software subsystems, namely communication (Wi-Fi, Bluetooth and mobile), sensors, power supply, the Linux kernel and Android's application framework. This section coarsely describes the contents of the data set resulting after the privacy-preserving measures. Table 1 summarizes global properties of the data set.

3.1   Contents

The data may be categorized into high-level user context, external sensing, and system internals.

Table 1. Dataset properties.

Content	Smartphone utilization, OS, and sensor data
Num. of participants	22
Num. of data points	2.8×10^8
Size	1.8 GB bzipped CSV
Timespan	4 months
Privacy measures	global/individual substitution

High-Level User Context is utilization data that contains direct hints to the participant's current activity and context. This includes the state of the display (on/off, brightness) and the phone (idle, ringing, or off the hook). Also the currently running packages belong to this category. Settings can also indirectly tell about the participant's context. For example, turning the ringer to silent mode, when it was set to play a ringtone before, is a hint that the situation changed to one that prohibits phone noise, like a meeting or a cinema. Besides audio settings, also the communication settings, whether Bluetooth or Wi-Fi is enabled, or whether the device is in airplane mode, belong to this category.

Sensing data is obtained from various physical sensors as well as positioning and communication hardware. The physical sensors measured acceleration, magnetic field strength, orientation and light intensity. When the participant allowed it, also information about current altitude and speed were obtained from the GPS hardware. Also, communication devices can be used to sense the presence or even the signal strength of (potential) peers. Whereas Wi-Fi and baseband processors deliver information about stationary communication peers (access points and cell towers) and are thus feasible for positioning, Bluetooth delivers information about mobile communication peers.

System Internal data mainly describes the overall usage of the system's resources like the CPU, the battery, the main memory and the network interfaces. The use of Android *wakelocks* also belongs to this category. Since wakelocks are used to prevent the device from suspending, (application) bugs regarding their handling can severely increase energy consumption.

3.2   Structure

For every device, our data set contains a stream of events. Every event is composed of a timestamp, an attribute name, and the new value for this attribute. Table 3 shows an excerpt of such a stream. For entity types with multiple instances, like Wi-Fi access points, attribute names contain a unique identifier for this resource (e.g., the BSSID for Wi-Fi). The appearance and disappearance of such entities is denoted with "1" or "0" respectively. For example, at time 1346837529394, package "vax98" is started, whereas at 1346837579524, the access point "kcz63" has gotten out of reach. The complete data set consists of 280 million of these events.

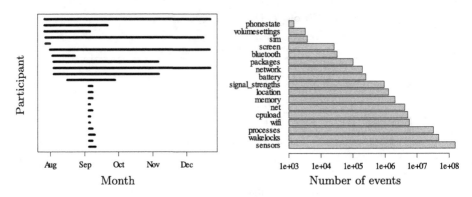

Fig. 1. Left: Period of participation for every participant. Right: Total number of events for every event type (changes only, log scale).

Figure 1 illustrates their distribution regarding event type and participant. Table 2 contains all attributes in condensed form. A detailed and exhaustive description can be found at the data set's website.

4   Analysis

Many different kinds of analysis can be imagined on the data set presented here. Among them semantic place prediction [9], network cell prediction [4], transportation mode detection [10], frequent subsequence mining as well as the prediction of resource consumption [11,12]. Due to the streaming nature of our data set, the streams abstraction [13] was used for preprocessing[3]. In the following, we show how the data can be incorporated for smartphones power modeling [11], generative modeling of application usage and a network cell prediction task [4].

4.1   Energy Model

Researchers have proposed a number of power models for ubiquitous systems [11,12,14]. Usually, power models are derived manually by using a power meter attached to one specific system instance. As a result of the model derivation process, the generated power model is at best accurate for one type of ubiquitous system and at worst accurate only for the specific system instance for which it was built. It would require great effort and time to manually generate power models for the wide range of phones now available.

We now show how a simple linear regression power model

$$\hat{y}_t = \beta_0 + \boldsymbol{x}_t \boldsymbol{\beta}$$

---

[3] The stream container and processors that have been written to preprocess the data for both tasks are available online at: http://sfb876.tu-dortmund.de/mobidata.

Table 2. Condensed names of collected attributes including the data **C**ategory (**H**igh-Level, **S**ensing, or **I**nternal), the sampling **I**nterval, and the data **S**ource (Android **API** or Linux **VFS**). The sampling interval is either given in seconds or as the fact that we received the data as callback event (**E**) whenever it changed. Values in square brackets are placeholders for actual identifiers. An asterisk means that the according value (or identifier if in the bracket) was replaced for privacy.

Attributes	Cat.	Int.	Src.
airplanemode, phonestate	H	2,E	A
battery:{health, level, plugged, status, temperature, voltage}	I	E	A
bluetooth:{, connected:[*mac_address**], device:[*mac_address**]: {, name*, class, bondstate, prev_bondstate, rssi}}	H,S		A
cpuload:{1min, 5min}	I	2,60	V
location:network:{accuracy, time, location:gps:{accuracy, altitude, speed, time}	S	60	A
memory:{Buffers, Cached, Dirty, MemFree, MemTotal, Writeback}	I	60	V
net:wifi:{, {r, t}x_{bytes, packets}}, net:mobile:{, {r, t}x_{bytes, packets, dropped, errors}, multicast}	I	60	V
network:{roaming, cell:{cid*, lac*}, operatorid*, type}	S	2	A
notifications:vibrate, ringer:{silent, vibrate, volume}, media:volume	H	2	A
packages:{launchable:[*package**], running:[*package**]}	H	60	A
processes:[*pid*]:{, cmdline:{*, parameters*}, state, tcomm*, {u, s, cu, cs}time, priority, nice, num_threads, start_time}	I	60	V
screen:{brightness, on, timeout}	H	2	A
self:{skip, start}			
sensors:{azimuth, light, pitch, roll, {x,y,z}force, {x,y,z}Mag}	S	120	A
signal_strenghts:{gsm_signal_strength, gsm_bit_error_rate, cdma_dbm, cdma_ecio, evdo_dbm, evdo_ecio}	S	E	A
sim:{state, serial*, subscriberid*}	I	300	A
wakelocks:[*name*]:{active_since, {, expire_, wake_,}count, last_change, {max, sleep, total}_time}	I	2	V
wifi:{, connection:{bssid*, link_speed, network_id, rssi, ssid*, supplicant_state}, scan:[*bssid**]:{, capabilities, frequency, level, ssid*}}	H,S	60	A

can be estimated with our data set, whereby we follow the approach that was presented by Zhang et al. [11]. The event stream is converted to a set of consecutive windows. Since the energy consumption should be predicted, we consider the change of battery level as label, i.e. $y_t = \text{batLevel}_t - \text{batLevel}_{t-1}$. The following measurements are considered as features: mobile network and Wi-Fi signal strength, Wi-Fi speed, number of outgoing/incoming Wi-Fi and mobile network packets in the last window, display brightness, GPS usage and CPU utilization.

Table 3. Example data for one device.

Timestamp	Attribute	Value	Timestamp	Attribute	Value
1346837529316	network:cell:cid	kjx37	1346837529469	cpu:load:5min	3.49
1346837529346	wifi	True	1346837529512	network:operatorid	xoz28
1346837529366	wifi:connection:ssid	qlk83	1346837529633	net:wlan0:tx_bytes	31342252
1346837529394	wifi:scan:kcz63:ssid	dle45	1346837530254	packages:running:vax98	1
1346837529428	ringer:vibrate	True	1346837530317	sim:serial	ldm52
1346837529451	screen:on	False	1346837530351	phone:state	idle
1346837529454	screen:brightness	100	1346837534507	bluetooth:devices:jtm18	1
1346837529468	battery:level	39	1346837579524	wifi:scan:kcz63	0

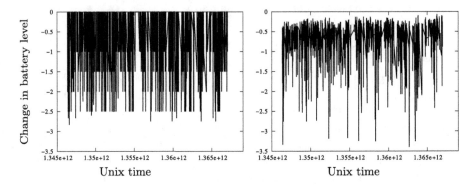

Fig. 2. Energy consumption of a smartphone as measured by the change in battery level over time. Left: measured consumption. Right: predicted energy consumption.

Figure 2 shows the measured energy consumption for one user over time on the left, and the corresponding prediction on the right. Let $\boldsymbol{y} = (y_1, y_2, \ldots, y_N)$ be the vector of measured values and $\hat{\boldsymbol{y}} = (\hat{y}_1, \hat{y}_2, \ldots, \hat{y}_N)$ a vector that contains predictions. The 10-fold cross validated root mean squared error

$$\text{RMSE}_N(\hat{\boldsymbol{y}}, \boldsymbol{y}) = \sqrt{\frac{1}{N} \sum_{t=1}^{N} (\hat{y}_t - y_t)^2}$$

of the estimated linear model is 0.604 with a deviation of ± 0.02 and the mean absolute error

$$\text{MAE}_N(\hat{\boldsymbol{y}}, \boldsymbol{y}) = \frac{1}{N} \sum_{t=1}^{N} |\hat{y}_t - y_t|$$

is 0.525 (± 0.013), both for a 10 min window width. Thus, the predicted drop in battery level is on average about 0.52 % off.

If the future energy consumption has to be predicted with such a linear model, predictions of the future feature values are required. One approach is to build a generative model for each of the features. Due to the different types of data (like the number of sent network packages or CPU utilization), many different models have to be learned. Joint models of variables with different types of marginal distributions are very resource-intensive in terms of computational power, since they often require Markov-Chain Monte-Carlo (MCMC) sampling procedures for training and prediction. Energy models that rely only on information about running applications (apps) and location information as features instead of mobile network, display, CPU, Wi-Fi and GPS usage are cheaper to compute.

4.2   Application Usage

Now, we show how a generative model of application usage can be constructed, that might serve as an input for other discriminative procedures like energy models. The utilization of n applications over time generates a sequence or stream of binary vectors $\boldsymbol{x}_t \in \{0,1\}^n$, whereas

$$\boldsymbol{x}_{t,i} = 1 \Leftrightarrow \text{App } i \text{ is running at time } t.$$

In general, points in time t might correspond to a time interval like $t \hat{=} [8\text{:}00\text{am}; 9\text{:}00\text{am}]$. In this case $\boldsymbol{x}_{t,i} = 1$ iff app i is running between 8:00am and 9:00am. The vectors $\boldsymbol{x}_t$ are interpreted as realizations of a multi-variate random variable $\boldsymbol{X}_t$ which is itself part of the multi-variate random variable $\boldsymbol{X} = (\boldsymbol{X}_t)_{1 \leq t \leq T}$. $\boldsymbol{X}$ represents the full time period of length T that has to be modeled, like $\overline{T} \hat{=}$ "one day" or $T \hat{=}$ "one week". Here, we model the application usage over one day. Together with a time interval width of 60min, we have $T = 24$.

If we recall how we use our private smartphone over the day, it appears that certain apps might prohibit or imply the use of other apps. Playing a game prevents the usage of other apps, while web browser and e-mail client are likely to run at the same time. To detect and extract such dependencies, the low-order conditional independence graph $G_0 = (V_0, E_0)$ [15] is estimated on a sample of the data stream.

Model. Based on this graph, a Spatio-Temporal Random Field (STRF) [16] is constructed. STRF is a special kind of undirected probabilistic graphical model with compressed parameters for spatio-temporal data. Here, we interpret the low-order conditional independence graph G_0 as spatial graph. This graph is copied for each point in time $t = 1, \ldots, T$. The copies or *layers* are interconnected to form a temporal chain $G_1 - G_2 - \cdots - G_T$. The connections are given by a set of spatio-temporal edges $E_{t-1;t} \subset V_{t-1} \times V_t$ for $t = 2, \ldots, T$ with $E_{0;1} = \emptyset$, that represent dependencies between adjacent graphs G_{t-1} and G_t, assuming a Markov property among layers, so that $E_{t;t+h} = \emptyset$ whenever $h > 1$ for any t.

The resulting spatio-temporal graph G, consists of the vertex set $V := \cup_{t=1}^T V_t$ and the spatio-temporal edges connecting them $E := \cup_{t=1}^T E_{t-1;t}$. This is in contrast to the graphical structure from [16], where the spatial edges of each layer E_t were included in the edge set of the spatio-temporal graph G. Including the spatial edges in a STRF model is motivated by the fact that real world sensors might fail. In case of a sensor failure, knowledge about the values that are measured at neighboring sensors at the same time can help to reconstruct the missing values. However, this kind of failure can not happen in the app data—we will always know all apps that are running "right now" at time t and hence, spatial edges that help to impute missing values are not needed.

Finally, G is used to induce a generative probabilistic graphical model that allows us to predict (an approximation to) each apps maximum-a-posteriori (MAP) state as well as the corresponding marginal probabilities. The full joint probability mass function of the STRF is given by

$$p_{\boldsymbol{\Delta}}(\boldsymbol{X} = \boldsymbol{x}) = \frac{1}{Z(\boldsymbol{\Delta})} \prod_{v \in V} \psi_v(\boldsymbol{x}) \prod_{(v,w) \in E} \psi_{(v,w)}(\boldsymbol{x}),$$

with sparse parameters $\boldsymbol{\Delta}$, partition function Z and potential function ψ. By construction, a single vertex v corresponds to a single app a at a fixed point in time t that can either run or not. The potential function of an STRF has a special form that obeys the smooth temporal dynamics inherent in spatio-temporal data.

$$\psi_v(\boldsymbol{x}) = \psi_{a(t)}(\boldsymbol{x}) = \exp\left\langle \sum_{i=1}^{t} \frac{1}{t-i+1} \boldsymbol{\Delta}_{a,i}, \phi_{a(t)}(\boldsymbol{x}) \right\rangle$$

The STRF is therefore parametrized by the vectors $\boldsymbol{\Delta}_{a,i}$ that store a weight for each of the two possible values (running or not) for each app a and point in time $1 \leq i \leq T$. The function $\phi_{a(t)}$ generates an indicator vector that contains exactly one 1 at the position of the state that is assigned to an app a at time t in $\boldsymbol{x}$ and zero otherwise. Similar functions $\phi_{a(t),b(t')}$ that encode the edge states as 4-dimensional vectors exists for each edge $(a(t), b(t')) \in E$. For a given data set, the parameters $\boldsymbol{\Delta}$ are fitted by regularized maximum-likelihood estimation (MLE). Further details about the STRF model and the estimation procedure can be found in [16].

A convenient property of MLE with this kind of probabilistic model is the aggregated representation of training data. Instead of storing a set of observed vectors, MLE training of STRF requires the empirical expectation $\tilde{\mathbb{E}}[\phi(\boldsymbol{x})]$ which is a mean vector that can be computed directly on the data stream. This is especially well suited for resource constrained, ubiquitous systems.

Due to the continuous utilization of the device, training data (empirical expectation) is updated frequently. The model can be updated by re-starting the parameter estimation from the last known solution $\boldsymbol{\Delta}$. Re-training and prediction phases can be restricted to times when the battery is charged. The prediction for the current day can be precompute, based on the currently running apps. The operation system or other apps could then use the predicted information by looking-up the predicted $\boldsymbol{x}_{t,i}^*$ from memory.

Predictions are computed via MAP estimation,

$$\boldsymbol{x}^* = \arg\max_{\boldsymbol{x}_{V \setminus U}} p_{\boldsymbol{\Delta}}(\boldsymbol{x}_{V \setminus U} \mid \boldsymbol{x}_U),$$

where $U \subset V$ is a set of spatio-temporal vertices with known values. The nodes in U are termed observed nodes. Notice that $U = \emptyset$ is a perfectly valid choice that yields the most probable state for each node, given no observed nodes. To compute x^*, the sum-product algorithm [17] is applied, often referred to as loopy belief propagation (LBP).

Table 4. Properties of the application data and the network cell identifier data.

	User A	User G	User J
Number of windows	1202	1605	4574
Number of applications	86	58	86
Number of app edges	201	18	34
Number of network cells	70	48	37
Number of cell edges	258	65	9

Evaluation. Now, the quality of the spatio-temporal random field for application usage is assessed. Each participants data (Table 4) is splitted into two parts, a *training set* and a *test set*. Structural dependencies between apps are detected with $K = \frac{1}{4}$ (see [15, Proposition 6]). The model, as described above, is trained for each user with 150 iterations of unregularized maximum-likelihood estimation. For the quasi-newton optimization procedure, the curvature estimates are restricted to $D_{ii}^{-1} \in [10^{-6}; 10^0]$. Messages of LBP are damped by 0.5 to enforce convergence. LBP was stopped after 200 iterations in non-convergent cases.

For each time point in test set, we compute the prediction accuracy as the normalized number of correct predictions. Furthermore, we want to evaluate how frequently new predictions are required. This arises from the fact that STRF always predicts all layers, i.e. the complete day. A prediction that is based on the running apps at time t, can be used for the rest of the day and therefore, resources can be saved.

Figure 3 shows how the prediction performance for three participants varies over the day. In the left plot, a bar at time t corresponds to the average performance of a MAP prediction conditioned on the apps that were observed at time t. The prediction accuracy is computed for the layers $t + 1, t + 2, \ldots, T$ and averaged. This was repeated for each test instance and averaged again. The outcome of this procedure determines the height of the bar. In the right plot, the performance in predicting the next layer $t + 1$ is reported. Incorporating the most recent observations into the prediction does lead to the best results, but for certain kinds of applications of this model, it might be unnecessary to compute a new prediction every time a new layer is observed. In case of user A, the observations from 19:00 are sufficient to predict 90.0 % of the apps that run at the rest of the day correctly. Figure 4 shows a more extreme case: The predictions are just the unconditional MAP states, i.e. they are based on no observation at

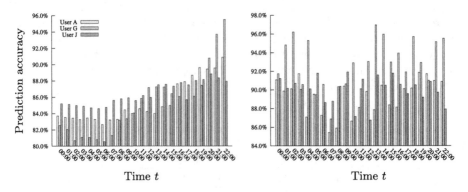

Fig. 3. Average conditional MAP prediction accuracy of the STRF for three participants over all test instances for 24 h. Left: Average over times $t + 1, t + 1, \ldots, T$, given running apps at time t. Right: Prediction accuracy for running apps at time $t + 1$, given running apps at time t.

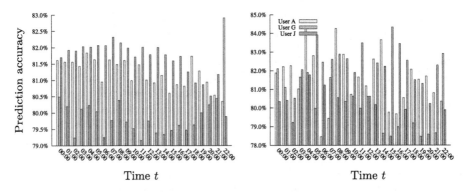

Fig. 4. Average MAP prediction accuracy of the STRF for three participants over all test instances for 24 h. Left: Average over times $t + 1, t + 1, \ldots, T$, given $\emptyset$. Right: Prediction accuracy for running apps at time $t + 1$, given $\emptyset$.

all. This is the cheapest solution in terms of resources, since the new MAP has to be computed only after the model is re-trained. Compared to the conditioned MAP prediction, the loss in accuracy is between 2 % and 10 %. For real world applications, a mixed version can be considered, where a prediction based on recent observations is used whenever enough energy is available and otherwise, the prediction is taken from the unconditional MAP or the last prediction.

4.3 Network Cell Prediction

Every mobile network connection has a unique network cell identifier (CID). A-priori knowledge about cells that a user will visit in near future can deliver an indicator about upcoming changes in availability of resources like network connectivity or charging opportunities. For a given set of features at time t, the network cell prediction task consists of predicting the network cells that the corresponding user will visit at times $t + 1, t + 2, \ldots, T$. A similar task has been addressed by Michaelis et al. [4] with support vector machines and linear chain Markov random fields. In contrast to the STRF approach, only one cell is predicted per time window.

The sequence of visited cells is extracted for each user and a sliding window of 60 min width is applied to this sequence. Each window can be interpreted as binary vector $x_t \in \{0, 1\}^n$

$$x_{t,i} = 1 \Leftrightarrow \text{Cell } i \text{ is visited in time intervall } t$$

that indicate, which cells the user has visited within a particular hour. The period length T is again 24. Although it is appealing to use the "true" geographical neighborhood of each cell to derive the conditional independence structure, the physical shape of a network cell is highly dynamic and multiple cells are likely to overlap. Furthermore, without access to a special database, the CIDs cannot be resolved to geographical information like GPS location. Instead of the geographical neighborhood, we derive the low-order conditional independence graph with parameter $K = \frac{1}{2}$ for this experiment. The edges show between which cells a user travels. Since the number of known cells goes up to thousands for some participants, we consider only those cells that appear in at least 1 % of all training data points as vertex.

Based on this graph, a STRF model is trained for 50 iterations, since more iterations did not lead to better results in this case. The remaining settings are the same as in the previous app data experiment. Since most of the cells are not visited in each hour, the classes are highly imbalanced. Therefore, Figs. 5 and 6 show the accuracy only for class 1, i.e. the number of correctly classified visited cells, normalized by the total number of visited cells. If we consider both classes,

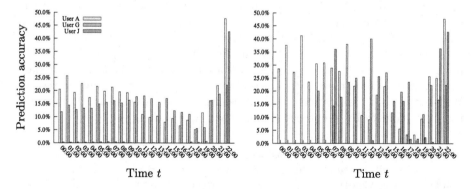

Fig. 5. Average conditional MAP prediction accuracy of the STRF for three participants over all test instances for 24 h. Left: Average over times $t + 1, t + 1, \ldots, T$, given running apps at time t. Right: Prediction accuracy for network cells at time $t+1$, given the cells that the user visited at time t.

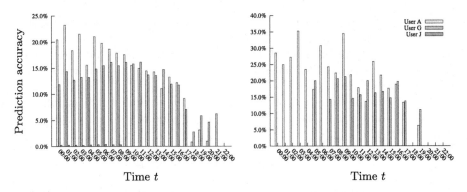

Fig. 6. Average MAP prediction accuracy of the STRF for three participants over all test instances for 24 h. Left: Average over times $t + 1, t + 1, \ldots, T$, given $\emptyset$. Right: Prediction accuracy for network cells at time $t + 1$, given $\emptyset$.

namely cells that are correctly classified as visited and cells that are correctly classified as not-visited, the overall prediction accuracy raises to 90 %.

In general, the accuracy is rather low compared to the previous experiment. Nevertheless, in case of user A, the model often predicts at least one of the visited cells correctly, which might be enough for a particular task. As expected, the conditioned MAP achieves a higher accuracy. For the prediction of cells that will be visited in the next time point (Fig. 5, right), the improvement of frequent re-predictions over unconditioned MAP predictions (Fig. 6, right) is about 5 % in terms of average prediction accuracy.

5   Conclusion

We presented our open smartphone utilization data set collected within our collaborative research center and during its summer school and presented some analysis to show its utility. We believe that open datasets greatly help to evaluate and improve analysis methods like these. It is interesting to see that, the further down the software-hardware stack a data source resides, the more data is generated and the more data is actually needed to obtain meaningful results. We see this as an indication towards the need for data collection frameworks that allow for flexible preprocessing and data aggregation.

Acknowledgments. This work has been supported by Deutsche Forschungsgemeinschaft (DFG) within the Collaborative Research Center SFB 876 "Providing Information by Resource-Constrained Data Analysis", project A1. We would also like to thank our collaboration partners from the EcoSense project at Aarhus University for providing technical support. Last but not least, we would like to thank all our participants for contributing to the data set.

References

1. Eagle, N., Pentland, A.: Reality mining: sensing complex social systems. Pers. Ubiquitous Comput. **10**(4), 255–268 (2006)
2. Kiukkonen, N., Blom, J., Dousse, O., Gatica-Perez, D., Laurila, J.: Towards rich mobile phone datasets: Lausanne data collection campaign. In: Proceedings of the 7th International Conference on Pervasive Services, ACM (2010)
3. Laurila, J.K., Gatica-Perez, D., Aad, I., Blom, J., Bornet, O., Do, T.-M.-T., Dousse, O., Eberle, J., Miettinen, M.: The mobile data challenge: big data for mobile computing research. In: Mobile Data Challenge by Nokia Workshop, in Conjunction with International Conference on Pervasive Computing, June 2012
4. Michaelis, S., Piatkowski, N., Morik, K.: Predicting next network cell IDs for moving users with discriminative and generative models. In: Mobile Data Challenge by Nokia Workshop in Conjunction with International Conference on Pervasive Computing, June 2012
5. Chon, Y., Talipov, E., Shin, H., Cha, H.: Mobility prediction-based smartphone energy optimization for everyday location monitoring. In: Proceedings of the 9th ACM Conference on Embedded Networked Sensor Systems, SenSys 2011, pp. 82–95. ACM, New York (2011)
6. Nath, S.: ACE: exploiting correlation for energy-efficient and continuous context sensing. In: Proceedings of the 10th International Conference on Mobile Systems, Applications, and Services, MobiSys 2012, pp. 29–42. ACM, New York (2012)
7. Schulman, A., Navda, V., Ramjee, R., Spring, N., Deshpande, P., Grunewald, C., Jain, K., Padmanabhan, V.N.: Bartendr: A practical approach to energy-aware cellular data scheduling. In: Proceedings of the Sixteenth Annual International Conference on Mobile Computing and Networking, MobiCom 2010, pp. 85–96. ACM, New York (2010)
8. Fricke, P., Jungermann, F., Morik, K., Piatkowski, N., Spinczyk, O., Stolpe, M., Streicher, J.: Towards adjusting mobile devices to user's behaviour. In: Atzmueller, M., Hotho, A., Strohmaier, M., Chin, A. (eds.) MUSE/MSM 2010. LNCS, vol. 6904, pp. 99–118. Springer, Heidelberg (2011)
9. Huang, C.M., Ying, J.J.-C., Tseng, V.: Mining users' behavior and environment for semantic place prediction. In: Mobile Data Challenge by Nokia Workshop in Conjunction with International Conference on Pervasive Computing, June 2012
10. Stenneth, L., Wolfson, O., Yu, P.S., Xu, B.: Transportation mode detection using mobile phones and GIS information. In: Proceedings of the 19th ACM SIGSPATIAL International Conference on Advances in Geographic Information Systems, GIS 2011, pp. 54–63. ACM, New York (2011)
11. Zhang, L., Tiwana, B., Qian, Z., Wang, Z., Dick, R.P., Mao, Z.M., Yang, L.: Accurate online power estimation and automatic battery behavior based power model generation for smartphones. In: Proceedings of the 8th IEEE/ACM/IFIP International Conference on Hardware/Software Codesign and System Synthesis, CODES/ISSS 2010, pp. 105–114. ACM, New York (2010)
12. Dong, M., Zhong, L.: Self-constructive high-rate system energy modeling for battery-powered mobile systems. In: Proceedings of the 9th International Conference on Mobile Systems, Applications, and Services, MobiSys 2011, pp. 335–348. ACM, New York (2011)
13. Bockermann, C., Blom, H.: The streams framework. Technical report 5, TU Dortmund University, December 2012

14. Kjærgaard, M.B., Blunck, H.: Unsupervised power profiling for mobile devices. In: Puiatti, A., Gu, T. (eds.) MobiQuitous 2011. LNICST, vol. 104, pp. 138–149. Springer, Heidelberg (2012)
15. Wille, A., Bühlmann, P.: Low-order conditional independence graphs for inferring genetic networks. Stat. Appl. Genet. Mol. Bio. **5**, 1–32 (2006)
16. Piatkowski, N., Lee, S., Morik, K.: Spatio-temporal random fields: compressible representation and distributed estimation. Mach. Learn. **93**(1), 115–139 (2013)
17. Kschischang, F., Frey, B., Loeliger, H.A.: Factor graphs and the sum-product algorithm. IEEE Trans. Inf. Theory **47**(2), 498–519 (2001)

Predictability Analysis of Aperiodic and Periodic Model for Long-Term Human Mobility Using Ambient Sensors

Danaipat Sodkomkham[1]([⊠]), Roberto Legaspi[2], Ken-ichi Fukui[1], Koichi Moriyama[1], Satoshi Kurihara[1], and Masayuki Numao[1]

[1] Institute of Scientific and Industrial Research, Osaka University, Osaka, Japan
danaipat@ai.sanken.osaka-u.ac.jp
[2] Research Organization of Information and Systems, Transdisciplinary Research Integration Center, The Institute of Statistical Mathematics, Tokyo, Japan

Abstract. The predictive technique proposed in this project was initially designed for an indoor smart environment wherein intrusive tracking techniques, such as cameras, mobile phones, and GPS tracking systems, could not be appropriately utilized. Instead, we installed simple motion detection sensors in various areas of the experimental space and observed movements. However, the data collected cannot provide as much information about human mobility as data from a GPS or mobile phone. In this paper, we conducted an exhaustive analysis to determine the predictability of future mobility of people using only this limited dataset. Furthermore, we proposed an aperiodic and periodic predictive technique for long-term human mobility prediction that works well with our limited dataset. The evaluation of the dataset collected of the movement and daily activity in the smart space for three months shows that our model is able to predict future mobility and activities of participants in the smart environment setting with high accuracy – even for a month in advance.

Keywords: Human mobility · Smart environment · Long-term prediction · Fano's inequality · Predictability analysis

1 Introduction

Understanding and predicting human mobility are crucial components in a number of real world applications. We will mention a few examples here. For example, the PUCK architecture [1,2], which automatically recognizes habitual activities, was introduced in a smart environments to provide reminders to people if they forgot important tasks, such as taking a prescribed medicine after a meal. This could be helpful for people who have dementia or mild cognitive impairment. Moreover, the ability to predict people's future locations is an important element in transportation planning [3,4], bandwidth provisioning in wireless local area networks [5], and targeted advertisement dissemination [6].

© Springer International Publishing Switzerland 2015
M. Atzmueller et al. (Eds.): MUSE/MSM 2013, LNAI 8940, pp. 131–149, 2015.
DOI: 10.1007/978-3-319-14723-9_8

In our specific example, we used an actual office environment with various built-in sensors and actuators to enable the pervasive computing technology to control different settings in the environment. Our prototype smart office environment was initially designed to create a working environment that can learn multi-users' behavioral activities and intelligently react to these activities smartly. Our objective in developing this smart environment is to simplify mundane repetitive tasks and to improve people's lives. For example, a smart office that could predict the future occupancy of a meeting room and automatically prepare the electronic facilities in the room prepared before the meeting. A smart office could also be programmed to predict needs from daily activities; for example, to ensure that hot coffee was ready to be served at a particular time. These applications require the ability to foresee individual's future whereabouts and mobility. This is referred to the open problem of long-term human mobility prediction.

In general, a smart environment has sensors that detect people's activities and mobilities. Different machine learning algorithms are then employed to explore meaningful information about behaviors and routines. This information will later be used to build a predictor that foresees people' needs and suggests a proper reaction. Therefore, a challenging problem for anyone researching smart environments would be *"to determine the best approach to observe people's mobility in the least obtrusive way, while providing enough information to build an accurate predictive model."*

There is an apparent trade-off between informativeness and conspicuousness of the sensing technique. For example, by using colored pictures from cameras with an image processing technique and semi-supervised classification algorithm, Yu *et al.* [7] were able to create a system that recognizes people and their positions. Moreover, they were directly able to map each individual's movement to a floor map. From this example, it is evident that there needs to be a balance between rich mobility information and discomfort caused by the cameras. Apart from the camera techniques [8,9] discussed earlier, mobile phone data [10,11], GPS [3,12], and RFID tagging [13,14] requires people to carry (or put on) a tracking device while in the environment, which is not feasible in real-world implementations. On the other hand, simple sensors, such as infrared distance sensors, ultrasonic distance sensors, and magnetic sensors, are small enough to blend into the environment and seamlessly observe human mobility. However, the data collected by simple sensors is less informative than high precision sensing technologies; as a result, the capability of the mobility predictive model built using simple sensors' data is limited.

In this paper, we investigate the limits of predictability for a mobility dataset derived from data collected by simple sensors. We also present a novel prediction technique, aperiodic and periodic (APP), for the long-term human mobility prediction problem. More specifically, APP technique predicts the future location of an individual at any specific *time frame* relatively far into the future, e.g. 21 days from the present between 10:00 and 10:05. The prediction is obtained by probabilistic models that compute how likely a certain location will be revisited in the future within the specific *time frame*. The prediction component in APP consists of two probabilistic models, periodic and aperiodic. Both models keep track of visited time stamps, extract contextual features from each visit (such as the time of day, the day of the week, or the duration since the last visit), and model their relationships. The *periodic model* is based on the hypothesis that a person periodically visits specific locations, such as every 3 h, every day, or every month. We analyze periodicity at each location and test this hypothesis. If the hypothesis holds, we use the periodic model; otherwise, the aperiodic model is used. The *aperiodic model* does not rely on periodicity; instead, it extracts significant patterns of repetitive movements representing past mobility behavior. The aperiodic model postulates that the same (or similar) mobility patterns tend to repeat at any specific future time frame whenever contextual features are similar. The combination these models into the APP technique results in a predictor that determines a person's location with acceptably high accuracy and precision – even a month in advance.

2   Limits of Predictability in Collective Human Mobility

Song *et al.* performed a breakthrough analysis of the predictability of human mobility [11]. They explored the limitation in predictability of an individual's movements. Note that they did not take the quality of the prediction techniques into consideration. Despite the differences in individual's daily behavior, their analysis over a large population monitored by mobile phone data shows 93 % potential predictability in an individual's mobility. In other words, predicting individual's movements can be effectively achieved when historical data is available.

When collective mobility data from multiple people rather than an individual person's historical data, the fundamental question of predictability for this class of data arises, i.e., *to what extent is collective mobility predictable?*

2.1   Collective Human Mobility Data

Our experimental smart environment was a functioning working space that included individual cubicle work stations, recreational space, and meeting areas. Twenty graduate students and faculty regularly work in this space. The floor plan is depicted in Fig. 1a. Individuals may have different duties, different class schedules, and different daily routines, which results in different directional and temporal mobility patterns. We installed two types of sensors in the environment to monitor activities and movements. Infrared distance sensors were primarily used to detect movement at each specific location. Magnetic sensors were attached to the hinge of the refrigerator and the oven to detect their use (Fig. 1b). All sensors were connected through our laboratory's network, and they continuously fed live streams of mobility data to a database. By employing these ambient sensors, participants did not need to be equipped with an intrusive tracking device during the experimental period and could move normally without being overtly aware that they were being monitored. We observed visitations and mobility in

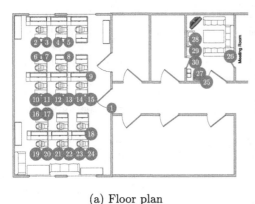

(a) Floor plan

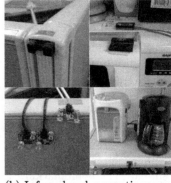

(b) Infrared and magnetic sensors used to observe mobility in the experimental space

Fig. 1. (a) Infrared and magnetic sensors used in the experiment (b) Placement of the sensors

the experimental environment 24 h a day for 3 months (precisely 92 days) during the autumn semester. The experimental space was divided into 30 locations of interest ($N = 30$); sensors were concealed at each located to record movements. We modeled human mobility with two representations for different purposes as follows.

Temporal sequences of repeated visitations. Collective mobility at each location is represented by the temporal sequence of repetitive visitations visited by unknown people during the observation. The state of visitation at a particular time is denoted by a binary value: 1 for *visited*, 0 for *not visited*. For instance, a sequence v_x represents mobility at location x from 00:00 to 23:59, with the sample rate μ of one sample per hour.

$$v_x = \left(\langle t'_0, 0 \rangle, \langle t'_1, 1 \rangle, \langle t'_2, 0 \rangle, \ldots, \langle t'_{23}, 1 \rangle \right),$$

where t'_i represents the observed *time frame* from $t_0 + i\mu$ to $t_0 + (i+1)\mu$, and t_0 is the start time, i.e., $t_0 = 00:00$ and $t'_0 = [00:00, 01:00)$.

Trajectories. By increasing the sensors' sample rate μ to one sample per 200 ms, we were able to record every visitation. Then, from a temporal sequence of visitations, $(\langle x_0, t_0 \rangle, \langle x_1, t_1 \rangle, \ldots, \langle x_{w-1}, t_{w-1} \rangle)$, we linearly searched for each transition point in the sequence where the transition time $t_{i+1} - t_i > 30$ s to create smaller sequences that represent trajectories.

Despite the unobtrusiveness and simplicity of the ambient sensing method, a considerable amount of the obtained data was noisy. To handle noises (such as false triggered events, sensors blocked by obstacles, and simultaneous trajectories from different people) and extract movement trajectories from the collective mobility dataset efficiently, we applied the sequential pattern mining algorithm PrefixSpan [15] to extract only sub-trajectories of length n that appeared in a set of all observed trajectories, $\mathbb{T}$, more frequently than a certain minimum number of occurrences, $support_{min}$, during the observation.

2.2   Limits of Predictability

We evaluated the predictability over the collective mobility dataset using the methodologies introduced by Song et al. [11]. By employing Fano's inequality [16,17], we assessed whether the upper limit of the probability of a moving person's destination could be correctly predicted given the most recent trajectory and the past collective mobility data.

Let T_i' denote a movement trajectory, and D_i be a destination of T_i' from the observations, $\mathbb{T} = (\langle T_0', D_0 \rangle, \langle T_1', D_1 \rangle, \ldots, \langle T_m', D_m \rangle)$. Given a predictor: $f : T_i' \mapsto D_i'$ that works well to predict a future location D_i of a moving individual based on recent length n movement trajectory T_i' and a set of length n trajectories $\mathbb{T}_n$, in which $\mathbb{T}_n \subseteq \mathbb{T}$, let e denote the event of failed prediction, i.e., $f(T_i') \neq D_i$, and let $P(e)$ be its probability. According to Fano's inequality, the lower bound on the error probability $P(e)$ can be found in the following inequality.

$$H(D|T') \leq H(e) + P(e) \log(N - 1). \tag{1}$$

Thus, the probability of predicting correctly, denoted by Π, is $1 - P(e)$. Namely,

$$H(D|T') \leq H(e) + (1 - \Pi) \log(N - 1), \tag{2}$$

where the destination D can take up to N possible locations and $H(e)$ is the corresponding binary entropy as follows:

$$\begin{aligned} H(e) &= -P(e) \log(P(e)) - (1 - P(e)) \log(1 - P(e)) \\ &= -(1 - \Pi) \log(1 - \Pi) - \Pi \log(\Pi). \end{aligned} \tag{3}$$

The conditional entropy $H(D|T')$ appeared in Eq. (2) quantifies the amount of information needed to predict the destination D, given recent trajectory T'. Given the probability $P(T')$ of the set of past trajectories $\mathbb{T}_n$ containing T' and the joint probability $P(T', d)$, the conditional entropy $H(D|T')$ is defined by

$$H(D|T') = \sum_{d \in D, T' \in T} P(T', d) \log \frac{P(T')}{P(T', d)}. \tag{4}$$

Then, we calculated the entropy $H(D|T')$ separately for each length n trajectories, i.e., $\mathbb{T}_n \subseteq \mathbb{T}$, and analyzed the maximum potential predictability (denoted by Π^{max}) or the probability of predicting the destination of a person correctly given the collective mobility dataset by solving for the Π^{max}, where $\Pi \leq \Pi^{max}$ in Eq. (5), according to Eqs. (2), (3) and (4)

$$H(D|T') = -(1-\Pi^{max}) \log(1-\Pi^{max}) - \Pi^{max} \log(\Pi^{max}) + (1-\Pi^{max}) \log(N-1). \tag{5}$$

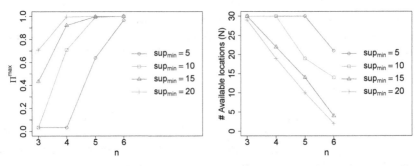

(a) The probability Π^{max} as functions of trajectory length n

(b) Relation between the parameter $support_{min}$, trajectory length n, and the number of predictable destinations N left after noise removal

Fig. 2. Predictability of collective human mobility in smart environment: Π^{max} is the upper bound of the probability that a particular predictive algorithm is able to predict a person's location correctly using only the collective dataset.

Figure 2a shows Π^{max} as functions of n, where n denotes length of the considered trajectories. It is not surprising that n increases predictability; a longer trajectory provides the predictor with more evidence, which helps narrow the search space. The $support_{min}$ also shows potential to eliminate unusual trajectories in the dataset, and gives significantly higher potential predictability. However, there is a trade-off between the degree of predictability and the number of predictable locations, as shown in Fig. 2b. A high threshold of minimum support ($support_{min}$) results in fewer numbers of locations N available to the predictive algorithm.

To summarize, despite the fact that the collective human mobility cumulative movements and behaviors from different people and seems diverge to the experimenter in the first place, accurate prediction of the location of a person is achievable with acceptably high probability. However, the analysis does not provide any clue about potential predictability for a long-term prediction configuration when the inference of a person's mobility cannot rely on recent movement patterns and frequent historical trajectories. Moreover, predictive techniques that work well for short-term human mobility predictions cannot be extended to long-term predictions effectively [12,18]. Therefore, in the next section, we studied the possibility to employ the periodicity in human behavior to foresee their mobility in far future instead of directly modeling trajectories.

3   Periodicity in Collective Human Mobility

Even without the use of data mining tools, it is evident that most of human activities are periodic to some extent. If a certain action or movement pattern is repeated regularly with a particular interval τ, and if this behavior is consistent over time, it is certainly predictable with the time period τ. In addition, the probability of predicting the correct location of an individual in the future depends on the tendency of such mobility patterns recurring at intervals. Therefore, we define *periodicity probability* to quantify this property formally.

Definition 1. *Let $P_x(\tau)$ denotes the periodicity, which is the probability of a particular event x reoccurring regularly with the constant time interval τ, where τ is a positive integer. Given the temporal sequence, as described in Sect. 2.1, of events from t'_0 to t'_m in which the location x was visited, the periodicity probability $P_x(\tau)$ is defined by*

$$P_x(\tau) = P\big(v_x(t'_{i+\tau}) = 1 | v_x(t'_i) = 1\big),\ t'_i \in \{t'_0, t'_1, \ldots, t'_{m-1}\}, \tag{6}$$

where $v_x(t'_i)$ indicates the state of the visitation at x during the time frame t'_i.

At the location x_1, apparent daily periodic behavior is revealed in the density plot presented in Fig. 3a, where dense areas are concentrated and aligned on a certain *time of the day*. In contrast, the density plot of x_2 in presented Fig. 3b shows slightly weaker daily periodicity. The loosely dense areas that are distributed broadly over time suggest a low degree of certainty of the repetition. However,

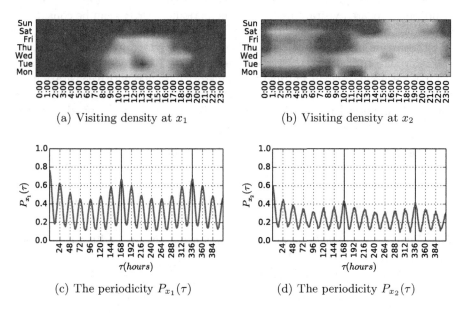

(a) Visiting density at x_1 (b) Visiting density at x_2

(c) The periodicity $P_{x_1}(\tau)$ (d) The periodicity $P_{x_2}(\tau)$

Fig. 3. In (a) and (b), the density of visitations at location x_1 and x_2 related to *time of the day* and *day of the week* are depicted, respectively. Busy times and days, in which a high number of visitations occurred within the same period of time, are shown in dark red. Dark blue indicates the opposite. The periodicity $P_{x_1}(\tau)$ and $P_{x_2}(\tau)$ of the corresponding locations x_1 and x_2 are shown as a function of time period τ, in (c) and (d), respectively.

weekly periodic behavior is evident in x_2, as distinct shades appeared on each the *day of the week* row indicating unique behavioral patterns on each day.

To find significant periodicity in the collective human mobility, we searched for τ that maximizes the periodicity probability at each location separately. Figure 3c and d show two sample locations where periodic behavior can be observed. The small peaks in these plots reveal relatively high probability that these particular locations were visited regularly with the time period τ, when τ is in multiples of 24 h. Moreover, the maximum predictability probabilities are found at multiples of 168 h. Without doubt, this clearly indicates that *daily* and *weekly* behaviors exist in the collective mobility data. Using a more algorithmic method to find significant period τ, the Fourier analysis also suggested that $\tau = 24$ h and 168 h correspond to two of the most significant frequencies of $\approx 4.167 \times 10^{-2}$ Hz and $\approx 5.925 \times 10^{-2}$ Hz, respectively.

In the next section, we analyze the possibility of the collective human mobility being predictable with periodic behavioral patterns.

4   Predictability of the Periodic Model

Intuitively, the periodicity $P_x(\tau)$ can be assumed to have estimated the precision of a periodic-based predictive model, which is based on a strong assumption of periodically repeated visitations. Hence, the periodicity $P_x(\tau)$ can be considered as a measurement for the predictability of the periodic model. In addition, we want to provide another predictability analysis applying an academic concept from information theory to the periodic model.

First, we assign *periodic entropy* to the history data of repetitive visitations at each location to determine the amount of information needed to foresee future visits given historical records of repetitive visitations. At each location x, the periodic entropy is computed as follows.

Definition 2. *Given the collective mobility data, the entropy S_x^τ that quantifies the degree of uncertainty of the periodicity $P_x(\tau)$ in the dataset is as follows:*

$$S_x^\tau = \sum_{\nu \in \{0,1\}} P(v_x) H(v_x(t'_{i+\tau}|v_x(t'_i) = \nu)), \quad t'_i \in \{t'_0, \ldots, t'_{m-1}\}, \qquad (7)$$

where $P(v_x)$ is the probability of a location x being visited, and the conditional entropy $H(v_x(t'_{i+\tau})|v_x(t'_i) = \nu)$ is

$$H(v_x(t'_{i+\tau})|v_x(t'_i) = \nu) = \sum_{\varphi \in \{0,1\}} P(\varphi|\nu) \log\left(\frac{1}{P(\varphi|\nu)}\right), \qquad (8)$$

where $P(\varphi|\nu)$ stands for $P(v_x(t'_{i+\tau}) = \varphi|v_x(t'_i) = \nu)$.

In addition, let S_{xf}^τ be the entropy of future visitations; i.e.,

$$S_{xf}^\tau = -\sum_{\varphi \in \{0,1\}} P(v_x(t'_{i+\tau}) = \varphi) \log(P(v_x(t'_{i+\tau}) = \varphi)), \quad t'_i \in \{t'_0, \ldots, t'_{m-1}\} \quad (9)$$

Next, we determine the predictability for each location x of the periodic model with the probability $\Pi_{x,\tau}$ defined as follows.

Definition 3. *Let $\Pi_{x,\tau}$ be the probability that the periodic model predicts times of future visitations at x correctly by always predict visits at all times that are $k\tau$ apart from the last visit for $k = 1, 2, \ldots$ Thus the associated entropy $H(\Pi_{x,\tau})$ of the predictability $\Pi_{x,\tau}$ is as follows:*

$$H(\Pi_{x,\tau}) = -\Pi_{x,\tau} \log_2(\Pi_{x,\tau}) - (1 - \Pi_{x,\tau}) \log_2(1 - \Pi_{x,\tau}). \tag{10}$$

The maximum predictability $\Pi_{x,\tau}^{max}$ can be determined using Fano's inequality in accordance with Eq. (2).

$$S_x^\tau \le H(\Pi_{x,\tau}) + (1 - \Pi_{x,\tau}) \log_2(N - 1). \tag{11}$$

Because $\Pi_{x,\tau} \le \Pi_{x,\tau}^{max}$ and $N = 2$ prevents this boundary to the binary classification, the following correction is required.

$$\begin{aligned} S_x^\tau &\le H(\Pi_{x,\tau}) + (1 - \Pi_{x,\tau}) \log_2(N - 1) \le H(\Pi_{x,\tau}) + (1 - \Pi_{x,\tau}) \log_2(N) \\ &= -\Pi_{x,\tau}^{max} \log_2(\Pi_{x,\tau}^{max}) - (1 - \Pi_{x,\tau}^{max}) \log_2(1 - \Pi_{x,\tau}^{max}) + (1 - \Pi_{x,\tau}) \log_2(N). \end{aligned} \tag{12}$$

After solving for $\Pi_{x,\tau}^{max}$ in Eq. (12), the predictability $\Pi_{x,\tau}^{max}$ determines the upper limit of the probability of predicting future visits of people at location x in the far future given an appropriate periodic model (with the time period τ). We evaluated S^τ and Π_τ^{max} separately for each location, and the associated distribution of Π_τ^{max} is shown in Fig. 4a. Both distributions of the predictability Π_τ^{max} indicate the average predictability over all locations is approximately greater than 80 %, in both daily and weekly models. The average predictability of the weekly model is slightly higher and has lower variance than the daily model. It

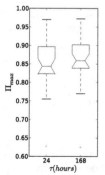

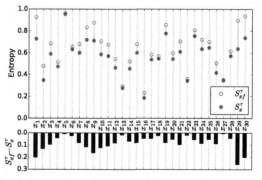

(a) Predictability of periodic models.

(b) Differences between the periodic entropy and the entropy of future visits at each location.

Fig. 4. Predictability Π_τ^{max} and corresponding periodic entropy

is reasonable to conclude that the weekly model fits the collective mobility data better than the daily periodic model.

Figure 4b shows differences between the periodic entropy, S_x^τ, and the entropy of future visitations, S_{xf}^τ, at each location x when $\tau = 24\,\mathrm{h}$. Note that as S_x^τ is closer to zero and farther from S_{xf}^τ, future visitations are more likely to periodically depend on previous visitations. For instance, locations x_5, x_6, x_{17}, x_{18}, and x_{28}[1] were not periodic, while the locations $x_1, x_2, x_8, x_9, x_{29}$ and x_{30} appeared to be more periodic than others. Therefore, the periodicity-based predictive model alone would not work in all locations; hence, we have developed the integrated aperiodic and periodic model for long-term human mobility prediction.

5   APP: Aperiodic and Periodic Model for Long-Term Human Mobility Prediction

APP, our long-term human mobility predictive model, combines two paradigms. The first approach (periodic approach) employs the periodic property in human mobility to foresee future visits. The second approach (aperiodic approach) does not rely on the periodicity; rather, it presumes that mobility patterns are similar to a day in the past that has similar features. APP model uses one of the two approaches to predict human mobility at a certain location x depending on the periodicity probability $P_x(\tau)$ at that specific location. If $P_x(\tau)$ is more than the user-specific threshold P_{min}^τ, then APP uses the periodic approach. Otherwise, it switches to the aperiodic approach.

5.1   The Periodic Approach

APP model's periodic predictive approach was designed to foresee times of future visitations at each location in the smart space. To predict future locations of multiple people, the predictions are computed independently for each location. Then all the results are combined to provide a set of locations that are likely to be visited at the specific time in the far future.

The fundamental idea behind the prediction is based on the assumption of periodicity. If the visitations at x recur regularly and repeatedly with constant time interval τ and if this periodic behavior appears consistently over time, then the probability $P_x^\tau(t_f')$ that the future visitation will occur within the time frame t_f', when the last visit happened at t_{m-1}' can be computed as

$$P_x^\tau(t') = P(v_x(t_{m-1-(k)\tau+\delta}') = 1 | v_x(t_{m-1-(k+1)\tau+\delta}') = 1), \ k = 1, 2, \ldots, \lfloor m/\tau \rfloor \tag{13}$$

where $\delta = \tau - (f - m + 1) \bmod \tau$.

This simple, yet accurate, predictive approach works well only at locations, where individuals' mobility has apparent periodicity. Otherwise, the periodic approach obtains poor predictions because mobility in those particular locations is not governed by periodic behavior. To address this problem, we proposed the optional aperiodic approach, which is independent of the periodic behavior.

---

[1] Placement of each point of interest x_i can be found in Fig. 1a.

5.2   The Aperiodic Approach

In the second predictive technique implemented in APP method, we extract significant patterns of repetitive visitations that occurred on different days at each location. Next the days that have a similar visiting behavior pattern are clustered, resulting in groups of *similar days*. Then, we extract contextual features from each group of similar days. Due to the limited dataset available, in this project only two features of interest were considered: (1) day of the week, (2) whether or not it was a holiday. Note that unlimited additional features, such as temperature, traffic, weather conditions, or meeting schedule, that might relate to the mobility pattern can be used to more comprehensively characterize the day.

The intuition that supports this predictive approach is derived from the weekly model presented in Sect. 5.1, i.e., human mobility patterns on the same *day of the week* are likely similar. In addition, human activities on national holidays are apparently different from normal workdays; therefore, we need an additional bit to explicitly specify this property. Hence, the mobility pattern of *a day* can be individually modeled by the visitations at each location. Recall the temporal sequence v_x described in Sect. 2.1; mobility at a certain location x can be represented by a vector:

$$d_x = [v_x, day_{week}, holiday]$$
$$= [v_{t'_0}, \ldots, v_{t'_{23}}, Sun, Mon, \ldots, Sat, Hol]. \tag{14}$$

The day vector d_x consists of 32 bits. The first 24 bits model visitations at location x during a specific time frame of a day, which is divided into hours. The next 7 bits indicate day of week, and the last bit indicates a holiday.

Similarity between two day vectors is generally measured by the Hamming distance [19]. Then, the k-means clustering algorithm [20] is applied to a set of day vectors to find clusters of similar days. Note that the clustering considers only the first 24 bits, the visitation part of day vectors. The parameter k of the algorithm directly implies the number of different mobility patterns that occurred on different days. The centroid of each cluster now represents a common mobility pattern that provides predictions (probability) of visitations on particular future days that have similar features. A concrete example of similar day clusters from the real dataset is shown in Fig. 5. The cluster centroids in Fig. 5 clearly show three different visiting patterns at that particular location. Cluster (1) contains a set of days in the past history when visitations rarely

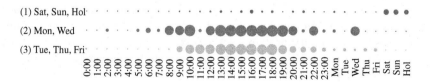

Fig. 5. Three cluster centroids that represent three mobility patterns.

Table 1. Human Mobility Dataset

Dataset	Training	Test
Sample rate	hourly	
Number of participants	≤ 20	
Observed locations	30	
Size of data	62 days (1,488 h)	30 days (720 h)

happened, and the majority of this set are Saturdays, Sundays, and days specified as holidays. On the other hand, clusters (2) and (3) contain more active days. The days in cluster (2), primarily Mondays and Wednesdays, have very low visitations records from 11.00–12.00 and 21.00–22.00; moreover, the visitations seem to occur earlier than on the days in cluster (3). Interestingly, this follows from the fact that we have scheduled meetings in the experimental space every Monday and Wednesday, which causes the mobility pattern on these days to be different from other days.

6   Evaluating Prediction Performance

In this section, we report on an evaluation of the prediction performance of the proposed long-term human mobility predictor on a physical collective human mobility dataset accumulated inside the working environment. As described in Sect. 2.1, the dataset contains 92 days of mutual movements of all participants. Data were collected consecutively 24 h a day, 7 days a week from approximately 20 participants using infrared sensors and magnetic sensors. These sensors were installed at 30 locations over the experimental space to detect activities and mobility at each area (Fig. 1a). Movements and activities were not scripted beforehand; all actions occurred spontaneously or deliberately in relation to each individual's routine, work schedule, and needs at that instant.

First, we evaluated the periodic approach for long-term human mobility prediction. Two months of collective mobility data was used to build the predictive model and the remaining 30 days of mobility data were used to test the model. Details of the dataset are summarized in Table 1.

6.1   Periodicity and Prediction Performance

We determined relations between the periodicity probability ($P_x(\tau = 24)$ and $P_x(\tau = 168)$) and the prediction accuracy precision and recall rate at each location separately. Figure 6a and d exhibit a decreasing trend of prediction accuracy with increasing periodicity probability; however, the periodic predictor returns higher precision and recall rates as the dataset has higher probability of such movements being periodically repeated. Nevertheless, the measurement of prediction accuracy is meaningless to us because the datasets, which contain historical visitations records for each location, have negative skew. In other words,

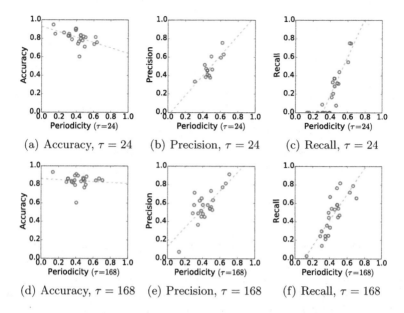

Fig. 6. Periodicity and prediction performance

a naive predictor could achieve at least a 60 % chance of correctly predicting visitations (either *"visited"* or *"not visited"*) at a specific time frame in the future by always guessing *"not visited"*. Figure 6b and e show a direct relationship between the precision rate and periodicity probability. Likewise, the recall rates in Fig. 6c and f show that the datasets with higher periodicity are more predictable than others. Moreover, when the periodicity probabilities are lower than 0.4, the daily periodic approach (Fig. 6c) clearly returns poor results, i.e., those visitations were not daily periodic and were hardly covered by the weekly periodic model. These results confirm our hypothesis that the periodic approach alone is not effective for predictions with low periodicity probability.

6.2 Prediction Performance of the Aperiodic Approach

The aperiodic part of APP is implemented with the similar-day predictive approach described in Sect. 5.2. In previous experiments, the periodic approach underperformed on the datasets where mobility was not really periodic. This was particularly the case for the daily periodic model (see Fig. 6a, b, and c when most locations in the experimental space had periodicity probability lower than 0.4. Hence, in this experiment, the aperiodic part of APP is activated when periodicity is lower than the minimum threshold $P_{min}^{\tau} = 0.4$, the specified threshold.

Figure 7 reveals the benefit of including the aperiodic component in the APP predictive model. In Fig. 7a, the precision rates of the APP model after the implementation of the similar-day approach for low-periodicity data are significantly improved compared with the periodic approach alone. The precision

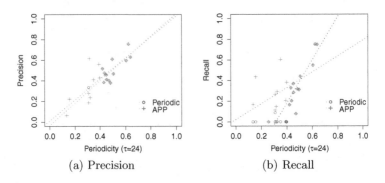

(a) Precision   (b) Recall

Fig. 7. Prediction performance of the similar-day approach

plots of the periodic approach on the left (periodicity ≤ 0.4) were mostly omitted from analysis because the periodic predictor never predicted *"visited"* for those locations, resulting in undefined precision rates.

APP also improves the recall rates, as shown in Fig. 7b. It is striking that the recall rates used to achieve close to 0.0 with the periodic approach increase to 0.6 when the APP predictive technique is employed. Samples of prediction evaluations of the two most accurate and the two least accurate predicted locations are shown in Fig. 8. In the most accurate case, APP was able to obtain the area under the receiver operating characteristic (ROC) of 0.79, where the sensitivity of the decision threshold was varied (Fig. 8b). Nonetheless, the worst predictor scores occurred for a ROC of 0.48.

In this final evaluation, we measured how close the predicted visitations are to the actual visits. The prediction error is simply the distance between the predicted timestamp of an expected future visit and the timestamp of the closest actual future visit. The results are summarized in Fig. 9. Impressively, 60 % of the estimation errors are less than 2.5 h (with mean = 7.5 h, and median = 1.5 h), considering that the prediction was made a month in advance.

In summary, the aperiodic part in our proposed long-term human mobility predictive technique improves prediction performance, particularly when the periodicity probability is too low to infer future visitations. However, the similar-day approach in the aperiodic part is not sufficiently effective to improve the predictive technique that employs the weekly periodic approach ($\tau = 168$). The reason for this is that the *day of week* feature resulted from cluster analysis in the similar-day approach corresponded to the weekly periodic model, and *holiday* is not a significant feature since there were few holidays during the three months when the dataset was collected. Implementation of the similar-day approach (aperiodic part) and the weekly periodic approach did not achieve significant improvement compared with implementation of only the weekly periodic predictive approach.

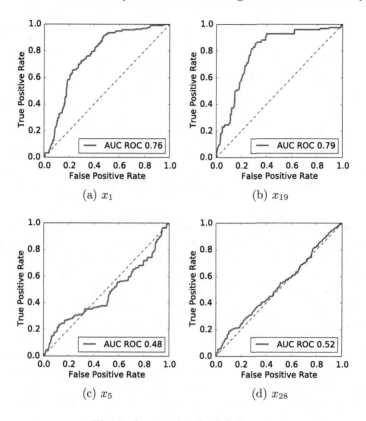

Fig. 8. Area under the ROC curve

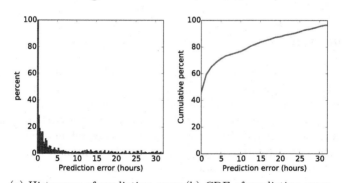

(a) Histogram of prediction error (b) CDF of prediction error

Fig. 9. Distribution of prediction error

6.3 Long-Term Prediction Performance

In this section, we report the results of testing the robustness of APP over a long-term. (Details of the dataset are summarized in Table 1.) Prediction performance

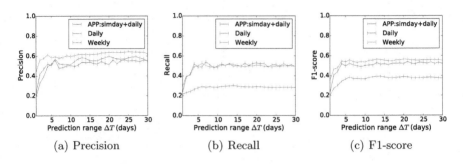

(a) Precision   (b) Recall   (c) F1-score

Fig. 10. Long-term prediction performance

for each day was summarized and plotted across a prediction range of 30 days. The results presented in Fig. 10 show steady prediction performances even when predicting for 30 days in the future. The F1-score, which is the harmonic mean of the precision and the recall rate (Fig. 10c), summarizes the prediction performance of the three proposed predictive techniques as follows. First, the collective mobility dataset that initially appears to be random contains sufficient information to enable accurate predictions even in the far future. Activities and corresponding mobility in the dataset are likely to be periodic on a weekly basis; hence, the weekly periodic predictive approach alone can achieve an average F1-score of 0.55. On the other hand, the daily model performs relatively poorly (average F1-score of 0.37) with this dataset because of the low periodicity probability on a daily basis. However, after implementing the similar-day approach together with the daily predictive model, the integrated model can achieve an average F1-score of 0.52.

7   Discussion

Human mobility and activities monitoring technologies have been improved notably in the past decade. Many advance non-intrusive techniques, such as WiFi signal strength-based techniques [21–23] and laser-based tracking [24], have been developed and made available for human behavior researches. Similarly, we aim at unobtrusiveness and simplicity and opt for the ambient sensors, in which tracking sensitivity is relatively higher than the other methods. In addition, the hierarchical probabilistic model and statistical method proposed in [25] are also capable of applying to non-intrusive sensing data and, yet, are able to recognize multi persons' behaviors. The limitation that these techniques are facing, however, is that they cannot distinguish individual identities. Consequently it is impossible to create an individual predictive model for each individual's mobility pattern. That is, despite advance acquisition techniques, the obtained data are still mixed up and inseparable among multi users.

Given such collective data, the predictability analysis revealed the potential for building a short-term next location predictive model that can accurately predict the next movement of a mobile individual using only the collective dataset (implementation of a short-term predictor is outside the scope of this paper). We also discovered acceptably high predictability for long-term prediction by modeling periodic behaviors hidden in the collective mobility data. The predictability analysis confirms that such diverse movement patterns from different persons are not entirely useless for predictive models training job. Although our conclusion was from one specific environment, the predictability estimation methods for both short-term and long term prediction can be generalized to justify other data acquisition techniques as well.

8   Conclusions

In this paper, the answer to the question about limitations of the predictability of collective human mobility from simple ambient sensor data collected in a smart environment has been found. Our decision to use non-intrusive tracking methods unavoidably caused some difficulties.

We proposed the APP: aperiodic and periodic predictive model for long-term human mobility prediction. Experimental results showed that the performance of the APP predictor significantly improved while predicting in low periodicity situations using the aperiodic approach. However, the aperiodic approach implemented in this project is not yet sufficiently effective to increase the performance of the weekly periodic predictive approach because very limited features can be extracted from the collective dataset. We intend to address this challenge in future studies.

Acknowledgement. This work was partly supported by JSPS Strategic Young Researcher Overseas Visits Program for Accelerating Brain Circulation and JSPS Core-to-Core Program, A. Advanced Research Networks.

References

1. Das, B., Chen, C., Dasgupta, N., Cook, D.J., Seelye, A.M.: Automated prompting in a smart home environment. In: Proceedings of the 2010 IEEE International Conference on Data Mining Workshops (ICDMW '10), pp. 1045–1052 (2010)
2. Das, B., Cook, D.J., Schmitter-Edgecombe, M., Seelye, A.M.: PUCK: an automated prompting system for smart environments: toward achieving automAted Prompting-challenges Involved. Pers. Ubiquitous Comput. **16**(7), 859–873 (2012)
3. Monreale, A., Pinelli, F., Trasarti, R., Giannotti, F.: WhereNext: a location predictor on trajectory pattern mining. In: Proceedings of the 15th ACM SIGKDD International Conference on Knowledge Discovery and Data Mining (KDD '09), pp. 637–646 (2009)
4. Krumm, J., Horvitz, E.: Predestination: inferring destinations from partial trajectories. In: Dourish, P., Friday, A. (eds.) UbiComp 2006. LNCS, vol. 4206, pp. 243–260. Springer, Heidelberg (2006)
5. Song, L., Deshpande, U., Kozat, U., Kotz, D., Jain, R.: Predictability of WLAN mobility and its effects on bandwidth provisioning. In: Proceedings of the 25th IEEE International Conference on Computer Communications (INFOCOM '06), pp. 1–13 (2006)

6. Haddadi, H., Hui, P., Brown, I.: MobiAd: private and scalable mobile advertising. In: Proceedings of the Fifth ACM International Workshop on Mobility in the Evolving Internet Architecture (MobiArch '10), pp. 33–38 (2010)
7. Yu, S.I., Yang, Y., Hauptmann, A.: Harry Potter's Marauder's Map: localizing and tracking multiple persons-of-interest by nonnegative discretization. In: Proceedings of the 2013 IEEE Conference on Computer Vision and Pattern Recognition (CVPR '13), pp. 3714–3720 (2013)
8. Roth, S.: Discrete-continuous optimization for multi-target tracking. In: Proceedings of the 2012 IEEE Conference on Computer Vision and Pattern Recognition (CVPR '12), pp. 1926–1933 (2012)
9. Beleznai, C., Schreiber, D., Rauter, M.: Pedestrian detection using GPU-accelerated multiple cue computation. In: Computer Vision and Pattern Recognition Workshops (CVPRW '11), pp. 58–65 (2011)
10. Gonzalez, M.C., Hidalgo, C.A., Barabasi, A.L.: Understanding individual human mobility patterns. Nature **453**(7196), 779–782 (2008)
11. Song, C., Qu, Z., Blumm, N., Barabási, A.L.: Limits of predictability in human mobility. Science **327**(5968), 1018–1021 (2010)
12. Scellato, S., Musolesi, M., Mascolo, C., Latora, V., Campbell, A.T.: NextPlace: a spatio-temporal prediction framework for pervasive systems. In: Lyons, K., Hightower, J., Huang, E.M. (eds.) Pervasive 2011. LNCS, vol. 6696, pp. 152–169. Springer, Heidelberg (2011)
13. Baeg, M., Park, J.H., Koh, J., Park, K.W., Baeg, M.H.: Building a smart home environment for service robots based on RFID and sensor networks. In: International Conference on Control, Automation and Systems (ICCAS '07), pp. 1078–1082 (2007)
14. Hussain, S., Schaffner, S., Moseychuck, D.: Applications of wireless sensor networks and RFID in a smart home environment. In: Proceedings of the 2009 Seventh Annual Communication Networks and Services Research Conference (CNSR '09), pp. 153–157 (2009)
15. Pei, J., Pinto, H., Chen, Q., Han, J., Mortazavi-Asl, B., Dayal, U., Hsu, M.C.: PrefixSpan: mining sequential patterns efficiently by prefix-projected pattern growth. In: Proceedings of the 17th International Conference on Data Engineering (ICDE '01), pp. 215–224 (2001)
16. Fano, R.M.: Transmission of information: a statistical theory of communications. Am. J. Phys. **29**, 793–794 (1961)
17. Navet, N., Chen, S.H.: On predictability and profitability: Would gp induced trading rules be sensitive to the observed entropy of time series? In: Brabazon, A., O'Neill, M. (eds.) Natural Computing in Computational Finance. SCI, vol. 100, pp. 197–210. Springer, Heidelberg (2008)
18. Sadilek, A., Krumm, J.: Far out: predicting long-term human mobility. In: Proceedings of the 26th AAAI Conference on Artificial Intelligence (AAAI '12) (2012)
19. Hamming, R.W.: Error detecting and error correcting codes. Bell Syst. Tech. J. **29**(2), 147–160 (1950)
20. MacQueen, J.: Some methods for classification and analysis of multivariate observations. In: Proceedings of the Fifth Berkeley Symposium on Mathematical Statistics and Probability, vol. 1, pp. 281–297 (1967)
21. Ferris, B., Fox, D., Lawrence, N.D.: Wifi-slam using gaussian process latent variable models. In: IJCAI, vol. 7, pp. 2480–2485 (2007)
22. Ferris, B., Haehnel, D., Fox, D.: Gaussian processes for signal strength-based location estimation. In: Proceedings of Robotics Science and Systems, Citeseer (2006)

23. Pu, Q., Gupta, S., Gollakota, S., Patel, S.: Whole-home gesture recognition using wireless signals. In: Proceedings of the 19th annual international conference on Mobile computing & networking, pp. 27–38, ACM (2013)
24. Cielniak, G., Bennewitz, M., Burgard, W.: Where is...? learning and utilizing motion patterns of persons with mobile robots. In: IJCAI, pp. 909–914 (2003)
25. Nguyen, N., Venkatesh, S., Bui, H.: Recognising behaviours of multiple people with hierarchical probabilistic model and statistical data association. In: BMVC 2006: Proceedings of the 17th British Machine Vision Conference, British Machine Vision Association, pp. 1239–1248 (2006)

Author Index

Printed in the United States
By Bookmasters